The Future of Electric Vehicles in India

A Consumer Preference Survey

The Future of Electric Vehicles in India

A Consumer Preference Survey

PROF. (DR.) NIRUPAMA PRAKASH
Director, Amity Institute of Social Sciences,
Amity University Sector 125, Noida, Uttar Pradesh 201303, India.

Ms. RASHMI KAPOOR
Research Associate,
Swinburne University of Technology, Hawthorn VIC 3122, Australia.

MR. YASHPAL MALIK
Research Associate, Amity Institute of Social Sciences,
Amity University Sector 125, Noida, Uttar Pradesh 201303, India.

PROF. (DR.) AJAY KAPOOR
Pro Vice Chancellor (International Research Engagement and Development),
Swinburne University of Technology, Hawthorn VIC 3122, Australia.

ZORBA BOOKS

Published in India by Zorba Books, 2016

Website: www.zorbabooks.com

Email: info@zorbabooks.com

ISBN Print Book - 978-93-85020-71-1

Zorba Books Pvt. Ltd. (opc)

Gurgaon, INDIA

Printed in India

Contents

Chapter 3 Consumer Perceptions

Acronyms Used in the Report

Acronym	Explanation
AFV	Alternative Fuel Vehicles
CNG	Compressed Natural Gas
CO	Carbon Monoxide
CO_2	Carbon Dioxide
EV	Electric Vehicle (Here used for Full Battery Electric Vehicle, often called BEV)
EVI	Electric Vehicles Initiative
GDP	Gross Domestic Product
HEV	Hybrid Electric Vehicle
IEA	International Energy Agency
INR	Indian Rupees
Lakh	1,00,000
LCVs	Light Commercial Vehicles
LNG	Liquefied Natural Gas
LPG	Liquefied Petroleum Gas
NEMMP	National Electric Mobility Mission Plan
NOx	Nitrogen Oxides
PEV	Plug-In Electric Vehicle (Includes both BEVs and PHEVs)
PHEV	Plug-In Hybrid Electric Vehicle
SMEV	Society of Manufacturers of Electric Vehicles
SPSS	Statistical Package For Social Science
SUV	Sports Utility Vehicles

Preface

This book is a detailed exploratory research study based on the project undertaken between August 2013 to Jan 2016 to understand consumer preferences for the use of electric vehicles in the backdrop of socio-economic variables such as gender, age, income, education and region. Collaborative work by researchers from Amity Institute of Social Sciences, Amity University, Noida, India and from Swinburne University of Technology, Hawthorn, Australia led to this book which traces the evolution of the electric vehicle market across the world, and in India as it prepares for take-off here. It deals with the future of electric vehicles from the consumer's point of view.

Policy makers, environmentalists, automobile companies, students and researchers will find answers to the questions they are asking while getting ready for the era of responsible energy usage. They will benefit from this book about what the Indian consumer wants from electric vehicles in the coming years, and how society can make a concerted effort to ensure that the optimistic forecasts about electric vehicles become a reality in the immediate future.

Acknowledgements

The authors express their gratitude to Dr. Ashok K. Chauhan, Hon'ble Founder President, Amity Education Group for constant encouragement on undertaking research and project work. The authors are indebted to Hon'ble Chancellor, Amity University, Uttar Pradesh, Dr. Atul Chauhan who is a source of inspiration for academic excellence. We acknowledge with respect Prof. (Dr.) Balvinder Shukla, Vice Chancellor, Amity University, UP for her constant support and guidance in the endeavor to explore new vistas of knowledge that would be beneficial to society.

We duly acknowledge Prof. Linda Kristjanson, Vice Chancellor, Swinburne University of Technology, Hawthorn, Australia for her support and encouragement in research initiatives.

The authors acknowledge the financial support by Auto CRC, Melbourne, Australia for smooth conduct of the project leading to joint international publication.

Chapter 1

Introduction

Two major problems currently confronting the world are the energy crisis and growing environmental pollution. Energy consumption in the world is increasing faster as compared to previous years. It is estimated that the existing petroleum oil and natural gas reserves will only last the next few decades. The transportation and decentralized power generation sector depend to a large extent upon petroleum products, particularly on petrol and diesel. The growth of the transportation sector is also increasing significantly especially because of growing use of technology in the automotive industry. Now, people are able to afford personal vehicles in developing and under-developed countries as well.

Thus, the demand for petroleum accelerates the demand for crude oil petroleum products as well as their cost. In pursuance of the main objective of the project which is to study consumer preferences in the selection of EVs, work on the research study was started with a thorough review of the literature. It has emerged from the existing literature reviews that demographic, socio-demographic and environment related variables influence the stated preferences for EVs. (Train 1986; Brownstone et. al. 2000; Potoglou and Kanaroglou 2008; Caulfield, Farrell & McMahon, 2010; Ziegler, 2012).

Some researchers have classified important attributes used by consumers for evaluating vehicles such as, "purchase price, fuel cost, range between refueling/recharging, availability of fuel/recharging opportunities, vehicle performance (e.g., acceleration, top speed), single versus multiplefuel capability, and environmental performance (e.g. vehicle emissions)" (Golob et. al. 1993; Greene 1996) Others have generalized these into fewer categories such as monetary, non-monetary, and environmental (Potoglou and Kanaroglou 2008; Adler et. al. 2003; Bunch et. al. 1993). Most of the studies (Brownstone et. al.. 2000; Gordon and Sarigöllü 1998; Golob et. al.. 1997; Bunch et. al.. 1993) have presented conditional, multinomial, or nested logit models for choosing EV. EVs included in these models comprise "electric vehicles, LPG vehicles, hybrid electric vehicles, CNG vehicles, methanol vehicles and unspecified alternative fuel vehicles". Few of the studies revealed that compared to traditional vehicles, a lower emission rate will increase the associated individual's chance of selecting EVs, suggesting that at least a niche market and environmentalists will accept the innovative attribute of EVs (Kahn 2006; Gulati & Kandlikar, 2010; Akerlof & Kranton 2000). Fuel availability and fuel flexibility will both have a positive effect on selecting EVs (Ramadhas 2011).

In a number of studies reviewed by Khan (1986), Madre (1990), Dargay and Dermot (1999), a range of variables was found which can have a considerable effect on aggregate demand of automobile ownership. The monetary variables include "income or discretionary income, income index, gross national product (GNP), per-capita income, and household income, related to cost (price, cost index of motoring, transportation consumer price index, and private transportation expenditure), related to land use (urbanized area and population density or proportion of population in urbanized area), demographic characteristics (number of employees, size of the household, proportion of population coming from specific age, and economically active population) and other variables (automobile stocks, annual transit trips per capita within the area)" {Khan 1986; Madre 1990; Dargay and Dermot 1999}.

Studies that specialize in the availability of an alternative substitute fuel for passenger vehicles and provide a quick summary of this standing (including limitations) of the main AET technologies have also been reviewed (Mokhtarian and Cao 2003; Andan and Arcier 1997). There are also studies dwelling on the environmental attribute which may be considered an edge that AETs have over conventional vehicles (Gulati & Kandlikar 2010; Pfaff, Chaudhuri & Nye, 2004; Kahn 2006). While the results of some studies signify that issues like convenience and personal costs take primacy over environmental and geopolitical concerns and that our society fails to progress social and environmental idealism (Augustine, Sandidge and Williams 2011); others put forward the view that consumer approval can be improved only through the various EV technologies which are expected to emerge commercially over the upcoming decades. Consumer acceptance can be increased only through initiatives to increase the knowledge and understanding of these vehicle fuel types, growing adverse effects of pollution on the environment and on health, increasing awareness to reduce pollution and the need for reducing our dependence on gasoline as an energy source (Nixon and Saphores 2011).

Some of the studies analyzed survey data on potential consumer demand for electric cars and other EVs (Beggs, Cardell & Hausman 1981), other studies done by Sandidge and Williams (2011) used various forecasting methods to focus on the technologies for future mainstream adoption which have the maximum potential. Methods used for forecasting include scanning, which includes perusing news websites and articles, speaking with relatives and friends, and attending/organizing relevant conferences to identify the present foremost automotive fuel technologies. General survey method was used to understand the interest level of consumers for forecasting the market for EV; which helped in understanding key influential factors for purchasing the vehicles. By using survey method, demographic details for the respondents were also better understood. Other methods used include expert interviews together with academia, forecasting & futuristic studies experts, government officials and EV industry players (Sandidge and Williams 2011).

Various influential characteristics which are derived from the above survey methods can boost the sales of electric vehicles in the coming years. These are some of the possible measures which can affect the sales of EVs –

1. **The cost of batteries is expected to drop significantly:** Tesla is working on a battery model which will reduce the cost of the EV batteries significantly. Presently the average cost of EV battery varies between $US250-400/kWh which is going to come down to $US100/kWh in the near future (Vorrath, 2015).

2. **Apple is likely to launch electric cars** – Apple is likely to launch an electric car in the near future. Morgan Stanley says that if Apple joins the EV game, it would transform the industry landscape considering Apple's scale, innovation and integration capability (Vorrath, 2015).

3. **Progress is expected to happen around the range/charging issues** – Another important barrier to EV mass uptake is that even after undertaking a lot of research and development for increasing the driving range of EVs, most electric vehicles are still unable to surpass 100 miles i.e. 160 km per charge. However, according to Morgan Stanley, Tesla and GM are planning to launch models with driving ranges of more than 200 miles (320

km). More progress will be made in the coming future about the range and charging issues (Vorrath, 2015).

1.1 Why transport is important to human beings

Wessel (2012) explains that transport is one of the critical components to fulfill the demands of our routine life. He also highlighted that transport related services are required at each stage of life because we cannot meet our life's requirements and desires from one location. We have two options, either we have to move ourselves to things or make sure things are moved to us. Transportation permits economic development to take place because it's a basic prerequisite for human beings (Wessel, 2012).

There are two economic theories by which the significance of transport systems can be analyzed. These levels are macro-economic and micro-economic which are explained below:

- **Macro-economic level** studies the large parameters which can affect the transportation systems. These are often linked to "output, employment, and income within a national economy". Transportation has a significant role in a nation's Gross Domestic Product and normally contributes "between 6% and 12% of the GDP" (Rodrigue & Notteboom, 2016).
- **Micro-economic level** studies the small factors such as production-related costs, prices etc., which can affect the transportation system. Rodrigue & Notteboom analyzed that "higher income levels are associated with a greater share of transportation in consumption expenses. Transportation accounts for (on an average) between 10% and 15% of household expenditures, while it accounts for around 4% of the costs of each unit of output in manufacturing, but this figure depends greatly on sub-sectors" (Rodrigue & Notteboom, 2016).

The sustainability of transportation systems can be ensured by measuring the effectiveness and efficiency of the transport system along with the environmental and climate impacts of such systems.

1.2 Issues affecting transport system

The impact of the transportation system on the atmosphere is significant and it contributes to approximately 20%-25% of global emissions. When fossil fuels are directly burnt, it produces the maximum emissions which are almost 97%

(Earthineer, 2016). Because of the rapid growth of the transportation sector, it is feared that contribution of emissions in the environment can increase in the near future, especially by the road transport system.

There are social costs which are linked to transportation systems, which include various factors such as "air pollution, road crashes, physical inactivity, time taken by the family while commuting and vulnerability to fuel price increases". Congestion of traffic is another factor which limits the timely delivery of various goods and services. These factors create a negative impact on those people who are planning to buy their own cars. (Earthineer, 2016).

Today, conventional planning of transport system struggles to improve the mobility of everything. The main purpose of the transport system is to access the various functions that include "work, education, goods and services, friends and family". There are many techniques available in the world by which traffic congestions can be managed and simultaneously the adverse impact on environment and society can be reduced. Communities play a significant role in ensuring the sustainability of their transportation system (Earthineer, 2016).

There are many problems which are prominent in urban transport, they are:

Traffic jams and parking difficulties: Rodrigue & Notteboom (2016) analyzed that wherever there is a population of 1 million, the cities start facing fully packed traffic jams, especially during office hours (Rodrigue & Notteboom, 2016).

Shortage of public transport: During the peak hours of traffic, most of the public transport is either over-utilized or under-utilized. When public transport is under-utilized, it makes the services financially unviable. (Rodrigue & Notteboom, 2016).

High maintenance costs: Cities whose infrastructure is old are facing the financial pressure of maintaining it. It becomes extremely difficult for them to upgrade the transport infrastructure which involves high costs. (Rodrigue & Notteboom, 2016).

Adverse impact on environment: Vehicular pollution has a serious impact on the environment in urban areas which leads to severe health issues and hampers quality of life (Rodrigue & Notteboom, 2016).

Increased accidents and safety related concerns: In developing countries, the number of accidents has increased because of increased traffic. The accidents also lead to recurring delays. Increase in traffic leads to discomfort in using streets (Rodrigue & Notteboom, 2016).

1.3 Vehicular pollution

Cao (2014) mentioned that around the world, pollution created by vehicles has become a serious issue in the last few years. It is also one of the reasons for global warming. Since the second half of the 19th century, it is estimated that the earth's temperature has increased by 0.6 degree Celsius and greenhouse gases such as carbon dioxide (CO_2) have contributed significantly to climate change. It has also been proved in many studies that the maximum carbon dioxide is produced by the U.S. which accounts for around 23% of the global CO_2. The U.S. population contributes 4.6% of the total world's population and contributes 32.5% in the world's GDP. It is envisaged that by "replacing gasoline-based vehicles", the U.S. transportation sector will reduce the CO_2 emissions in a significant way. (Cao, 2004).

Since 1990, the EV industry has gained popularity because of various reasons such as growing consumption of petroleum products and their adverse impact on health and climate change. To promote EV technology, billions of dollars have been invested in the EV industry. Some of the EV models are already on the roads of U.S. while some are in their prototype stage. EV industry has the potential to change the market for gasoline vehicles and is expected to gain significant share in the automobile market in the coming years. On the other hand, some of the models of EVs have been discontinued because they were not able to meet customer expectations and these models had issues related to their economic viability also (Prakash et. al., 2014).

The EV industry depends on various external and internal issues, for example "vehicle performance, their limitations, consumer preferences, the regulatory environment, the price and availability of gasoline". This has created limitations for assessing the potential future market for EV.

1.4 Types of alternative transport

There are many factors which helped EV in gaining global attention such as limited reserves of fossil fuel, growing environmental pollution and unpredictability of crude oil prices. In the book "Alternative Fuels for

Transportation", authors discussed the zero-emission plan by three ways, "electric vehicles, fuel cells, and hybrid vehicles". It was also discussed that electric vehicles and fuel cell vehicles will not be in a position to compete with the hybrid electric vehicles in the near future (Ramadhas, 2011).

From various technologies available in the transportation sector, it can be concluded that there are six types of vehicles available in the market which are summarized below:

- **Bio-fuel:** "Bio-fuel vehicles have an internal combustion engine that is adapted to run on a type of fuel derived from biomass. The two leading types of bio-fuel are bio alcohol (an alcohol—most commonly ethanol—produced mainly from sugar and starch crops) and biodiesel (diesel made from vegetable oils, animal fats or recycled greases)".
- **Flex-fuel:** "Flex-fuel vehicles utilize a multi-fuel internal combustion engine that can run on different fuels—usually gasoline and bio-fuel—that are mixed into the same tank and burned as a blend in the combustion chamber. Flex-fuel vehicles normally run on a mixture of either ethanol or methanol and gasoline".
- **Battery Electric:** "Battery electric vehicles (BEVs) are propelled by an electric motor which is powered by a large on-board battery unit. Current battery units have a limited capacity and must be frequently recharged at specialized charging stations. Charging stations can usually be installed in homes on a standard 120V or 240V setup. Electric vehicles do not require a tail pipe for exhaust and have negligible emissions".
- **Hybrid Electric:** "Hybrid Electric Vehicles (HEVs) combine an internal combustion engine with an electric motor and a larger-than-normal car battery. Many current models use the internal combustion engine and regenerative braking technology to recharge the battery, but some newer models also have plug-in capabilities similar to pure BEVs. HEVs have a longer range than pure BEVs and fewer emissions than a gasoline vehicle, but still use fossil fuels as an energy source".
- **Hydrogen:** "Hydrogen vehicles use hydrogen gas as their primary fuel source, which produces no harmful emissions when burned. Hydrogen fuel can be produced domestically from methane and other fossil fuels, and also from alternative sources such as wind, solar, and nuclear power. Currently there are two implementations of a hydrogen vehicle: fuel cell

vehicles, where hydrogen reacts with oxygen in a fuel cell to power an electric motor; and internal combustion engine vehicles, where a traditional combustion engine directly burns hydrogen fuel".

- **Natural gas:** "Natural gas vehicles use compressed natural gas (CNG) or liquefied natural gas (LNG) as their primary fuel source. Natural gas is a clean alternative to gasoline and is even cleaner than many currently available EVs. Natural gas vehicles are powered by an internal combustion engine similar to gasoline engines, and existing gasoline engines can be converted to use compressed natural gas. The natural gas available in the United States is primarily produced in North America".

1.5 Brief about electric vehicles (EV)

An electric vehicle (EV) uses one or more electric motors or traction motors for propulsion. The electricvehicle first came into the market in the mid-19th century but was not fully introduced to the market until 1990s at which time there was a rising concern about the environment. In 1997, EV1 was released by General Motors. The model was a two-seater car powered entirely by electricity with three hours recharging time. However, the product was not launched in the market by GM despite having multiple buying options from various buyers for EV1. Later GM destroyed almost all the 900 models of EV1 and only a few models were left for the museum. Similar to GM, Toyota had also developed an EV for lease in California in the mid-1990s. The RAV4-EV was the only popular model among a small segment of environment friendly vehicles but it was eventually discontinued as public concern over global warming and pollution from fossil fuels peaked only in the 2000s. In a survey conducted in the US in 2008, it was analyzed that 51% of the respondents were interested in buying a car which can reduce their emissions and fuel costs (Dalwadi, Zhang & Karantinos, 2011). This was considered to be the beginning of a new era of electric vehicles.

There are mainly three types of electric vehicles in the world, they are, "hybrid electric vehicles (HEV), plug-in hybrid electric vehicles (PHEV) and single fuel all-electric vehicles (ZEV)".

Hybrid electric vehicles (HEV)

Dalwadi, Zhang & Karantinos (2011) explain that HEVs are based on the gasoline engine technology and braking power to charge the car's Ni-Cad or Ni-MH batteries. Toyota's Prius is the best example of HEV. The gasoline engine has been used in Prius and the application of the brakes is done to charge its nickel metal hydride battery while the car was running. The transition from gas to electrical power managed by a computer system helped Prius obtain a better fuel mileage and lower emissions than any vehicles at that time. Initially, the Prius was in demand with a relatively small segment of environmentalist consumers. It was not until gas hit $4 per gallon in 2008 that the Prius saw an incredible increase in sales and Toyota ramped up production, almost 10 years after it was first introduced to the market. From 2001 to the end of 2009, Americans bought 1.6 million HEVs, out of which one million were Toyota's Prius (Dalwadi, Zhang & Karantinos, 2011).

Plug-in hybrid electric vehicles (PHEV)

PHEV models use much of the same technology as the HEV, running on either electricity or gasoline and transitioning seamlessly from one to the other. However, there are two major differences between PHEV and HEV. Firstly, it is the battery. PHEV uses a Li-Ion battery instead of the Ni-Cad and Ni-MH batteries that are used in HEVs. This would allow PHEV to have a greater range and speedier recharge. Secondly, it is the way that battery works. The batteries for PHEVs could be charged by plugging into an electrical outlet. It would allow PHEVs to run on electricity exclusively until the battery ran out of power. Most PHEV models used a traditional three-prong plug and could be fully charged in 6 to 8 hours using the110-volt line. Also, they could be charged faster using a 220-volt line or at some special charging station (Dalwadi, Zhang & Karantinos, 2011).

Full electric vehicles, zero emission vehicles (ZEV)

The full electric vehicle is a truly emissions-free vehicle. The ZEV is powered solely by electricity, with zero emissions at the tailpipe. While charging ZEVs with 110-volt outlets, it could take as long as 8 hours for some models, charge time depending on many factors and the time could be shortened for some models, especially with some new high-volt charging innovations. The biggest challenge for this currently most environmental-friendly car in the world is the driving range. Most of the affordable EVs in today's market are within 100 mile limit per charge. However, there are a few models which are developed by Tesla Motors with a better driving range of 200-300 km on a single charge. These models are, "Tesla Roadster, Model S and Model 3" ("Driving Range for the Model S Family", 2014).

Although it is usually longer than most of the PHEV, for example, as for Nissan Leaf (ZEV) it can run over 80 miles per charge while as for Volt (PHEV) it can only run 40 miles per charge, PHEV does have alternatives to power the engine after the battery dies while ZEV will have no other resources to run the engine. For producing the clean energy for EVs, the large power plants can be situated in areas where pollution does not affect health. Additionally, technology can be used to reduce the pollution in these large scale plants. It was found that there are many stakeholders involved in the EV process which are mentioned in the next section (Dalwadi, Zhang & Karantinos, 2011).

1.6 History of EV

Development of EV can be divided into three phases as mentioned below:

- The rise of EV (19th and 20th Century)
- Current Status of EV (Early 21st Century, from 2000 to 2015)
- Future of EV (2016 onwards)

1.6.1 The rise of EV

Electric powered motor vehicles were first introduced in the middle of the 19th century. In the 1890s, the sales ratio of electric automobiles (EVs) and gasoline vehicles was 10:1. EVs ruled the roads and even most of the showrooms. A few vehicle manufacturers, like Oldsmobile and Studebaker, formed EV organizations as well. Later, an exclusive showroom for EVs was also introduced in the market. Until 1900, electric vehicles also held the fastest speed record ("Electric Auto Association", 2016).

Initially, the production of electric vehicles was done through manual assembling. Later, in 1910, when the demand for these vehicles increased, then EV organizations had to do the production through mechanized assembly lines. By this innovation, the manufacturers who were financially strong were able to produce the EVs, and the rest were not able to produce them because they were not able to purchase vehicle components in high volumes and as a result, they vanished from the market subsequently. This reduced the market size for EVs. Another reason for the declining market share of EVs was the negligible availability of infrastructure outside the city which could support the consumers for undertaking inter-city travel. Petrol cars were becoming available though they required manual start of the engine. Later, the electric motor also known as a starter was introduced in gasoline vehicles, which boosted the sale of gasoline vehicles and further reduced the market share of EVs. Due to these factors, the production of electric vehicles was stopped by the end of World War 1. EV was only available for the niche market such as "taxis, trucks, delivery vans, and freight handlers" ("Electric Auto Association", 2016).

Between the late 1960s and early 1970s, the production of EVs was again resumed because of growing air pollution caused by fossil fuel vehicles. Later in the 1990s, some leading auto makers resumed manufacturing of EVs which was supported under "California's landmark zero-emission vehicle (ZEV) Mandate". This EV also had limitations because it was produced in small numbers and assembled manually like their initial prototypes ("Electric Auto Association", 2016).

1.6.1.1 The first electric cars

Porsche

A patent was registered under the name of "then-known Ferdinand Porsche" for its electric motor in 1896. It was believed that "the motor was a breakthrough by its simplicity and the absence of an unreliable transmission system". Porsche persisted in trusting in the capacity of the electric car and evolved the Lohner-Porsche ElectroMobil. This automobile achieved a pinnacle pace of 50 km/h, with a battery weight of 410 kg, which could last upto 50 km in a single charge. By using the "front wheel hub motors", this became the first vehicle in the world based on front wheel drive technology. Porsche also developed a four-wheel drive model with the name of Lohner, but this model required 1800 kg of batteries for reaching the satisfactory driving range. In the meantime, the combustion engine got more reliability and less maintenance which was also required to maintain such vehicles. This halted the research and development work for the electric cars ("History of the electric car | LosApos.com", 2016).

The Lohner-Porsche

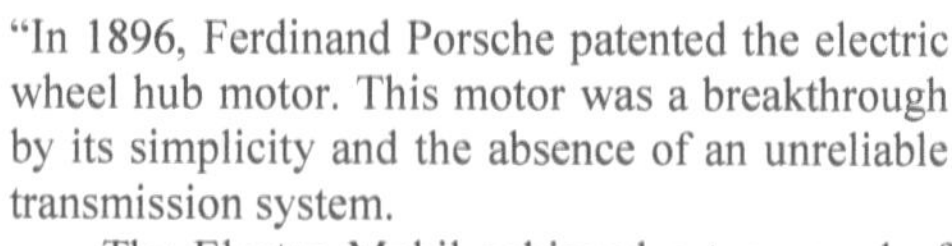

"In 1896, Ferdinand Porsche patented the electric wheel hub motor. This motor was a breakthrough by its simplicity and the absence of an unreliable transmission system.

The Electro Mobil achieved a top speed of 50 km/h, and with a battery pack of 410 kg, it had a range of 50 km.

With its front wheel hub motors, this was the first front wheel drive car in the world" ("History of the electric car | LosApos.com", 2016).

Thomas Edison and an electric car in 1913

"The American inventor Thomas Edison, inventor of the battery, and devotee of all things electric,including the electric car.

Here he is admiring the battery pack in his first car, a Baker Electric.

His friend Henry Ford consulted him about whether he should build the T-model with a petrol or an electric motor.

Unfortunately their discussions were not documented, but Henry's decision to go with petrol engines was to decide the future of the auto industry" (Sparke, n.d.)

1.6.1.2 The 1990s: Interest revival for EV

Despite remaining outside of the market for many decades, electric vehicles again got attention during the energy crisis in the 1970s and 1980s. In an auto show in Los Angeles during 1990, "General Motors President Roger Smith introduced the GM Impact electric concept car". A statement was also made during this event by GM that they would manufacture electric vehicles for the mass markets in the near future which would be available for sale as well ("Short History of Electric Vehicles - Anandalal Electric", 2016).

In the initial years of the 1990s, the California government through its board, known as the California Air Resources Board (CARB) started a discussion for fuel-efficient vehicles with a lower level of vehicular emission which can be also known as the zero-emission vehicle, i.e. electric vehicle. In addition to the initiative of CARB, various auto manufacturers started production of various EV models, such as, "the Chrysler TEVan, Ford Ranger EV pickup truck, GM EV1 and S10 EV pickup, Honda EV Plus hatchback,

Nissan lithium-battery Altra EV miniwagon and Toyota RAV4 EV" ("Short History of Electric Vehicles - Anandalal Electric", 2016).

During the 1990s, there was a substantial market developed for Sports Utility Vehicles (SUVs) which again lowered customers' interest in eco-friendly vehicles. These SUVs were economically unviable because they were not fuel efficient vehicles. Various American auto manufacturers concentrated on producing trucks as well which gave them higher margins in comparison to the small cars. These trucks were preferred in various places of Japan and Europe. Later, in 1999, the Hybrid car was launched by Honda in North America and it was the first vehicle of its type ("Short History of Electric Vehicles - Anandalal Electric", 2016).

Hybrid technology for an electric vehicle which was based on gasoline and electric power was a perfect match which offered eco-friendly and fuel-efficient vehicles without limiting the driving range. This vehicle was costly and could not get attention in initial years. The sales were poor because there was less interest in small sized cars and fuel-efficiency was not required at that time by most customers. Following the energy crisis which came in the 2000s, there was once again a boost to the EV market. As a result, Toyota Prius sales increased in the U.S. market. This encouraged other manufacturers to release their own hybrids or EVs. Many manufacturers started production of EVs because they felt that these cars would not be affected by oil price fluctuations in the future and would build a strong sales pipeline for them ("Short History of Electric Vehicles - Anandalal Electric", 2016).

1.7 Important stakeholders in the electric vehicle industry

There are several players in the EV's value chain: "Consumers, car dealers, and garages, gas stations, charging infrastructure manufacturers, battery manufacturers, car companies, utilities, financing and insurance companies and the government" (Dalwadi, Zhang & Karantinos, 2011).

1.7.1 Consumers

Consumers come at the end of the value chain but play acritical role. EVs can be purchased by individuals, corporations or fleet management companies but in any case, they look for similar attributes in their electric cars. The top attributes they look for in regular cars are economy and environmental friendliness. EV consumers' main concerns are mileage, especially for ZEVs, ease of use (charging considerations) and price (Dalwadi, Zhang & Karantinos, 2011).

1.7.2 Car dealers & garages

Car dealers and garages play an important role in making EVs available throughout the country. They will not show interest in stocking PHEVs and ZEVs unless they are certain that the necessary infrastructure i.e. charging stations, home-charging infrastructure etc., is in place to support the demand. Dealers are also concerned about the capital investment, workshop space, and the training that will be required to support EVs maintenance as they will have to move from repairing just ICE to repairing electric motors as well. Also, some characteristics of electric vehicles may render them less appealing to support: for example, the batteries are affected by extreme temperatures and this will reflect on how long the vehicle can stay at the dealer's shop. State regulations will play some role in the dealers "willingness" to adopt (Dalwadi, Zhang & Karantinos, 2011).

1.7.3 Gas stations – charging stations providers

Gas stations become important when it comes to charging either PHEVs or ZEVs. Currently, gas stations are the link that connects car users with the energy companies and, in that fashion; they can be the link that will connect electricity producers (practically the utilities) with electric car owners. In order to play that role, gas stations will have to invest in, currently, expensive quick charging infrastructure and, given that EVs are not abundant, retain some exclusivity over some geographic areas. On the other hand, as the driver will have to wait10 minutes for a full charge, providing charging services will allow gas stations to increase their high margin "mini mart" sales (Dalwadi, Zhang & Karantinos, 2011).

1.7.4 Charging infrastructure manufacturers

For the growth of EVs, charging infrastructure manufacturers will play a critical role. EV users will need to charge their cars and it is expected that they will be charging their EVs either at home or at work. There will be only a small number of customers who will go to charging stations to charge their vehicles. Charging infrastructure manufacturers are the link between the car manufacturers and the utilities, where they seem to have a two-way role: providing cars with electricity and the utilities with data for billing, forecasting etc. Charging infrastructure manufacturers are also linked to the battery manufacturers (how fast a battery can be charged is a function of both the charging infrastructure and the battery technology) (Dalwadi, Zhang & Karantinos, 2011).

1.7.5 Battery manufacturers

Battery manufacturers are the key stakeholders. The EV's battery performance is directly related to car price, performance, mileage, charging time and required equipment, as well as maintenance requirements. Also, because the cost of the car batteries is significant, various solutions have been proposed to affect the shape of the ZEVs' value chain; for example, the car companies will own the batteries and lease them to car owners or instead of charging the battery there will be battery-exchange stations (the latter has already been implemented in Israel). And those issues possibly affect the insurance and the financing of the EV purchase or lease. A key concern for battery manufacturers is whether new zero-emission technologies emerge and become commercially justifiable, such as the use of hydrogen fuel cells (Dalwadi, Zhang & Karantinos, 2011).

1.7.6 Car companies

Car companies are the main innovation drivers in the case of EVs. They have already invested significantly in the development of HEVs and some ZEVs and their growth in the future depends, partly at least, on the commercial success of the EVs. EVs can be the product through which car companies will extenuate their client relationships and develop new ones, among environmentally-conscious consumers. The widespread adoption of EVs will allow car manufacturers to move away from their "dependence" on oil companies. Furthermore, it will allow auto manufacturers to become players in the electricity market: not only will EVs consume significant energy, which means new business for electrical utilities, but also, through "smart grid" applications, the batteries of the electric vehicles can act as energy storage solutions that will allow utilities to stabilize demand and to better schedule electricity production (Dalwadi, Zhang & Karantinos, 2011).

1.7.7 Utilities

In EVs utilities will find a new source of demand for electricity as well as, with the development of the "smart grid", a way to reduce volatility in demand, as described above. Utilities will be responsible for bringing appropriate electrical power (some EV chargers operate on higher voltage than usual domestic outlets) to charging stations, which in some cases will require network upgrades etc. The "connecting link" between utilities and EVs are the charging stations which will have the "double role" of facilitating the charging, on the one hand, and the billing, data transmission etc. on the other (Dalwadi, Zhang & Karantinos, 2011).

1.7.8 Financing & insurance companies

Financing and insurance companies are important for the development of the EV market, as they already are for the ICE car market. At the moment, EVs command a price premium over comparable traditional cars and this makes prospective EV owners prime customers for financing providers. Furthermore, EV owners may need to finance the purchase or leasing of new batteries for their cars some years after they buy them and this also is an opportunity for financing companies. Similarly, insurance companies will need to innovate on offering insurance products that suit the characteristics of the EV and its use, e.g. insurance for charging station failure or damage. The challenge for both financing and insurance companies is that the business model for ZEVs is not clear yet, and itis not clear if more than one model will emerge, and this means that their tactics depend on those of other players (Dalwadi, Zhang & Karantinos, 2011).

1.7.9 Government

EV adoption is strongly encouraged by local, state and federal legislation. In the San Francisco Bay area there are in place three initiatives to support the adoption of EVs, sponsoring the use of an EV or the development of needed infrastructure: the Transportation Fund for Clean Air, the Alternative Vehicle, Infrastructure and the Innovative Climate Grants. At the Federal level, the American Recovery and Reinvestment Act provides $2.4 billion for EV-related infrastructure; furthermore, the acquisition of EVs is supported by tax credits ranging from $2,500 to $7,500 per vehicle (Dalwadi, Zhang & Karantinos, 2011).

1.8 Risks associated with EV

1.8.1 The charging infrastructure/gas station adoption risk:

The primary challenge in this ecosystem is who will initially bear the risk of implementing the infrastructure necessary to make EVs attractive to the mainstream consumer population. It is difficult to sell EVs in large volumes without a widespread charging infrastructure in place (which with current battery technology is necessary for trips over 40-100 miles) but in order to make investing in implementing a vast charging infrastructure attractive you need a critical mass of EVs.

Gas stations will play a critical role here. As electric battery performance, measured by mileage/charge and charging station performance, measured by charge time improve, existing gas stations can be retrofitted to include EV charging infrastructure with fewer investment costs. The key here is 1) lowering the investment cost required making the implementation more attractive and 2) providing a new stream of revenue to gas stations who currently operate on slim margins (especially on gasoline). Value-added services like streaming movies, music, Wi-Fi or payment processing might be opportunities for gas stations to improve margins and give them a reason to invest in EV charging infrastructure (Dalwadi, Zhang & Karantinos, 2011).

1.8.2 The consumer adoption risk

Another risk is associated with the battery portion of the value chain. As discussed earlier, there is the co-innovation risk, but there is also an adoption risk involved here. Battery makers have developed very sophisticated batteries capable of powering cars for substantial distances i.e. Tesla, Nissan Leaf (Dalwadi, Zhang & Karantinos, 2011).

Because of this lack of value, there is a large risk that customers will not adopt the current incarnation of EVs. The simple approach to address this adoption risk is to improve the batteries to the point that they create the same "value" as gasoline. The problem is that it will take time to advance battery technology to that point. To address this, the industry has taken on a hybrid approach. The Toyota Prius, one of the first commercially successful EVs, uses a hybrid engine that combines a primary combustion engine with an electrical system that can be used to supplement the combustion engine. GM, with their Chevy Volt, took a different track building a hybrid with its primary engine being electric and a combustion engine to provide power for longer trips. By doing this, the car manufacturers make the EV more attractive to mainstream customers. We believe it is a smart move because it is a step towards solving the chicken and the egg problem mentioned earlier. Car manufacturers are leading the way by taking the steps necessary to get EVs on the road with the hope that charging infrastructure will follow (Dalwadi, Zhang & Karantinos, 2011).

1.8.3 The utility electrical infrastructure adoption risk

The utility companies pose an interesting adoption risk. Because of heavy regulations, utilities measure success based on asset utilization and not pure profits. That generally implies that when they invest in infrastructure, they will attempt to draw every ounce of useful life out of it. What is interesting is that in the U.S., the power infrastructure is relatively old; the US power grid is sometimes described as a third world grid. So there are opposite forces pulling on the utilities. Firstly they want to get the most use of their current assets, but secondly they are facing rising maintenance costs of aging equipment and pressure from communities to update their power grids. The reason this is such a big deal is if EVs reach mainstream adoption, having hundreds of cars charging overnight in one neighborhood on one aging transformer will result in power grid failures (Dalwadi, Zhang & Karantinos, 2011).

The utilities so far have taken a wait and see approach. Utilities like Pacific Gas and Electric Company (PG&E), USA, do have incentive plans in place where EV owners get discounted rates when charging their vehicle at certain times during the day, but they have not invested a great deal in upgrading the power grid. Another interesting dynamic here that remains to be played out is the idea of a smart grid where customers can sell back electricity from solar cells or even a charged EV. Most likely there will be heavy government involvement here which most likely is also driving this wait and see approach by the major utilities (Dalwadi, Zhang & Karantinos, 2011).

1.8.4 Co-innovation risk

The success of the EV depends on many factors, e.g. the charging speed, the driving range per charge, the charging method etc. It is necessary to keep innovating around this electric power system. The most important co-innovation is obviously on the supply side, the battery and the charging device, as most of the consumers of EVs share the concern that the range that an EV can run per charge can put them in an awkward situation. Firstly, the range per charge is probably not long enough to cover the return trip. Secondly, the availability of charging places is way lower than the gas stations, making it harder to get the charging place. Even if the range is enough to cover the planned journey, an unexpected traffic event may put the driver in a very awkward situation. Thirdly, the charging times are too long. Unlike internal combustion engines (ICE) which usually take 10 minutes for a full refill that can support 400 miles, an electric engine will need 8 hours to recharge the car which can only run another 100 miles or less. Even fast charging systems will require a few hours of charging(Dalwadi, Zhang & Karantinos, 2011).

To solve this problem, the battery manufacturers will need to invent a battery that can give the electric engine longer hours to run and requires much less charging time. Also, the battery should be easy to be replaced or be charged by some easy carry-on electric device. Thus, when the battery runs flat, the driver will at least have some backup rather than waiting for someone to tow the car (Dalwadi, Zhang & Karantinos, 2011).

1.9 Influential attributes of electric vehicles

Electric transport is considered a mode of sustainable transport for most urban areas. There is a significant push from various key stakeholders, like government, industry players and societies in various countries to promote electric vehicles. In some markets, it has impacted the automotive market significantly and in some of the markets it is emerging as a key transport mode (Schulze, Müller & Meyer, 2015).

Nixon & Saphores (2011) discussed the models for preferences in their studies and found:

(1) Either the EVs are not available in the market;
(2) Or consumers are not able to get choices in the existing EV market.

By referring to substitutes available for gasoline vehicles, the consumers are not able to differentiate between the different characteristics of EVs. Hence, it is very difficult to come out with consumer preferences when the product itself is in hypothetical situations. To solve this, seven critical characteristics were identified to evaluate such vehicles, such as "price, fuel operating cost, range between refueling & recharging, availability of fuel/recharging opportunities, vehicle performance (e.g., top speed, acceleration), single versus multiple fuel capability, and eco-friendly performance (e.g., vehicle emissions)" (Nixon & Saphores, 2011).

Potoglou and Kanaroglou (2007) simplify these attributes into 3 main categories:

- Financial
- Non-financial, and
- Eco-friendly

1.9.1 Financial attributes

Nixon & Saphores (2011) explained that financial attributes are universally observed in "stated-preference studies" which include "vehicle buying price, fuel cost, and maintenance cost". There were few other benefits which were discussed such as "tax encouragements or sponsorships, free parking, and commute costs, involving access to express lanes". The purchase price for the vehicle and its on-going costs are the key factors which can drive the growth of EV industry (Nixon & Saphores, 2011).

It was mentioned by authors Gallagher & Muehlegger (2011) in their report that tax incentives provided by states are correlated with the adoption level of hybrid vehicles. Coefficients related to "income tax credits and sales tax waivers" were examined and it was concluded that incentives are linked to the sale of hybrid vehicles. Based on various estimations, it was concluded that "sales tax waiver of the mean value ($1,037) is associated with over thrice the effect of an income tax credit of mean value ($2,011)". Based on the values associated with incentives, it was estimated that waivers related to sales tax can boost the sale of hybrid vehicles by ten times. The authors also found that access related to "single-occupancy" is less correlated with adoption of EVs. However, prices of petrol are significantly correlated with the sale of hybrid vehicles. "For high fuel-economy hybrids, they have estimated that the cross-price elasticity of demand with respect to retail petrol price is 0.86". (Gallagher & Muehlegger, 2011).

1.9.2 Non-financial attributes

For EV, non-financial attributes such as "vehicle range between refueling (or recharging, in the case of electric vehicles), accessibility of fuel or recharging locations, and vehicle performance (e.g., acceleration, top speed)" are the reasons mentioned in most of the literature related to stated preferences. Other non-financial attributes are considered including "dual-fuel capability, refueling time, boot space limitations" for storage related issues, and the number of existing EVs in the consumer's geographical region. Findings suggested that for adopting EVs, range and kind of fuel availability are key limiting factors, after monetary concerns. Even though consumers understand the environmental benefits of EVs and generally have an optimistic attitude toward them, they are unwilling to give up average features of conventional vehicles (Ewing and Sarigöllü, 1998). Likewise, in the study of EV preferences among Southern California residents, it was noted that the importance of vehicle range for EV is significantly less than that of a conventional gasoline-powered vehicle (Bunch et. al.., 1993).

Cao (2004) highlighted that post the energy crisis of 1973 in the U.S., the government is putting in all efforts to control the energy consumed by the transportation department and has introduced several policies to reduce energy use. Some such initiatives were "reducing individuals' dependence on personal vehicles, and promoting higher average fuel vehicles". However, these policies were not very effective because there was a mismatch of needs among policy formulators and individuals. As a result, the sale of light duty trucks which includes SUVs and minivans increased post-1988. On the other side, the Energy Policy Act of 1992 encouraged the use of EVs to cut the dependence on oil import and preserve it for coming generations (Cao, 2004).

1.9.3 Eco-friendly attributes

Nixon & Saphores (2011) mentioned in their report that people are getting worried about growing air pollution and its impact on climate, especially because of burnt particles of fossil fuels. Electric vehicles have an advantage compared to conventional vehicles because they don't pollute the environment. Most of the studies related to the EV industry deal with the importance of eco-friendly attributes but provide negligible evidence which can influence the decision of the consumers. Most of the time, eco-friendly attributes are overshadowed by financial and non-financial attributes. However, a recent study involving drivers in Hamilton, Ontario (Canada) revealed that eco-friendly attributes are considered and do have a significant impact on their purchasing decision. It was revealed that "the likelihood of choosing an EV was found to be greater if pollution levels were 90 percent less than today's levels (significant at $p < 0.05$)". However, it is visualized that if electric vehicles reach 75 percent of the current emission level (Emission Level of 2011), the demand for hybrid or EV is expected to decrease in the future. So, caution will be required for doing a translation of coefficients for eco-friendly attributes, because social responsibilities might not influence customers to purchase eco-friendly vehicles. (Nixon & Saphores, 2011).

1.9.4 Socio-demographic characteristics

Nixon & Saphores (2011) highlighted that to get the working model for actual preferences of consumers, it becomes extremely critical to assess the effect of demographic preferences, socio-economic characteristics along with eco-friendly variables. In previous studies, it has been highlighted that males remain doubtful about adopting the electric vehicle because of certain limitations. They want to drive the vehicles with high speed even though they barely use the top speed of the vehicle during the vehicle's life. Similarly, aged people also remain doubtful about electric vehicles, however, educated people were found to favor EVs. Large families and longer commuters are found to be worried about the vehicle size and its driving range. From previous reports, it was analyzed that people who are concerned about environmental pollution are expected to buy the electric vehicle (Nixon & Saphores, 2011).

In their study "Car buyers and fuel economy", the authors Turrentine & Kurani (2007) tried to understand the thinking process and behavior pattern of U.S. customers for the automotive fuel industry. They interviewed fifty-seven families by using semi-structured questionnaires fornine lifestyle sectors and found that none of the families calculated their automobile cost in a systematic manner while making a purchase decision for their vehicle. Very few families were able to share the approximate cost for getting their fuel tank completely filled but they also could not share the accurate value. This explains that customers do not undertake in-depth analysis while purchasing a vehicle and end up purchasing a vehicle which is not economically viable in thelong run. Authors also found that the upfront cost of the vehicle is not the only cost which is given by the customers, the cost associated with the operating costs and fuel costs affects their budget in the long run. Many customers also consider the hypothetical value of fuel for future which is always on the higher side. (Turrentine & Kurani, 2007).

The authors analyzed several factors which are misperceived in the minds of customers while making a decision about fuel economy (Turrentine & Kurani, 2007). Some of the factors are mentioned below:

1. There are several vehicles in the market which can be differentiated by size, design, interiors, acceleration power, color patterns and lots of other attractive features, they are even equipped with the latest technology. With these features, customers forget to check the vehicle mileage and end up buying the vehicle without doing the cost-benefit analysis (Turrentine & Kurani, 2007).
2. In recent years, the reduced cost of oil has boosted the sale of conventional vehicles, however, there are many reasons which required attention of customers such as:
    - Uncertainty in the gasoline market in recent years.
    - Increasing global attention for reduction of CO_2 by the transportation sector.
    - Increased dependence on fuel imports which can depreciate the currency of a nation and adversely impact economy.

1.10 Global overview of electric vehicles

Almost 88 million passenger cars, buses, and trucks are manufactured every year worldwide ("Forbes Welcome", 2016). Most of these vehicles are dependent on fossil fuels such as petrol, diesel, CNG, LPG. In the future, air pollution, carbon emission, and perceived energy shortage are some of the factors which will increase the chances of electric vehicles of becoming the market leader. But there are some limitations of an electric vehicle such as long duration of charging time, battery maintenance, high purchase cost, short driving ranges and limited charging facilities which have become the roadblock for electric vehicle manufacturers (Perkowski, 2014).

Across the world, different approaches have been adopted to promote electric vehicles by many countries. Some offer investment support for technology, some offer infrastructural related investments and some give incentives to the buyers. As a result of such initiatives, 740000 battery operated electric cars were sold in the world market until December 2014 (Shahan, 2015). As a comparison, the total number of cars in the world is 1.2 billion (Voelcker, 2014). The market for EV is in the development process and has various limitations specifically related to the integration of various technologies, optimizing them and scaling up as per the needs of customers.

Some EV examples from various countries are mentioned below:

1.10.1 China

Various electric vehicle models such as Audi and BMW from Germany, Nissan and Toyota from Japan, Tesla and General Motors from the US, BYD and Roewe from China, were launched in 2014. In 2014, China manufactured 78,000 electric vehicles and 76,000 such vehicles were sold in the market, marking a growth of 350% & 320% respectively as compared to the previous year.

Growing demand for charging stations has attracted the government's attention as well. During 2010 to 2014, the number of charging stations in China jumped from 76 to 723 at the CAGR (Compound Annual Growth Rate) of 75.6% ("China Electric", 2015). For a comparison, there were approximately 99,000 filling stations in China till Dec 2014 ("China Filling", 2015).

1.10.2 Europe

Ernst & Young predicted that Europe will become the first mass market for electric cars by 2022. A survey of 300 European automotive leaders revealed that the industry is confident of growth over the next 12 months. Europe is expected to lead the way in the uptake of electric cars, China and Japan will occupy the 2nd and 3rd position in this market respectively. In 2015, more than 36,000 electric vehicles have been registered in the European market (Ayre, 2015).

1.10.3 Bhutan

The Bhutan government has set an ambitious goal to become the world leader by introducing a very large proportion of the government fleet of electric vehicles in their country. 50 electric vehicles of Nissan LEAF are on the roads and an order has been placed for another set of 22 such vehicles. So, in total, 77 EVs were sold in Bhutan and represents 10% of the total number of cars on Bhutan roads. EV fleet targets in Bhutan are police cars, protocol service cars, public transport in Thimphu, school buses, electric vans for transporting tourists, government ministry and the cabinet fleet (Tshering, 2014).

In line with these goals, in 2014, the Prime Minister of Bhutan introduced two models of electric vehicles –Reva from Mahindra and LEAF from Nissan, and in order to attract or win buyers, he suspended import duty as well. For example, Nissan offered a discount of almost 50 percent on the first 77 cars sold in Bhutan (each car sold for $14,516 each). Post completion of the offer, the vehicle will be sold at $28,000. The Bhutan government has planned to reduce the imports of fossil fuels to 70 percent by 2020 and the government does not want to promote economic development at the cost of its environment (Sundas, 2015).

1.10.4 India

The growth potential of electric/hybrid cars is yet to be realized in India, probably because they are considered to be not only expensive to buy, but also to maintain. The Indian government has also launched NEMMP 2020 to promote electric vehicles in the Indian market. If this plan is implemented successfully, then the Indian government will be able to save 9500 million tons of fossil fuel which is worth INR 62,000 crore ("National Electric", 2015). The Indian Government has planned to invest approximately INR 14,000 crores in the next five to six years with an additional investment of INR 8,000 crore, which will be pooled from auto makers ("In a Nutshell", 2013).

1.11 Indian government plans

There is a growing population in India, with the major growth in the large urban cities. This rising growth causes pollution concerns through an increase in the use of motorized vehicles within the urban environment. The large-scale urban migration of people from rural areas also puts high pressure on the infrastructure of large cities such as New Delhi, Mumbai, and Bengaluru. To reduce the pollution levels and the carbon emissions caused by the use of motorized vehicles, the government of India has initiated various programs to support the growth of EVs in the Indian market. Compared to the more developed solutions used by EU, US or Japan, Indian companies currently use low-tech solutions for E-mobility. There is a desire to develop a domestic market for design and manufacture of EVs and sub-system components by the Indian government (Global Opportunities for Electric Mobility: India, 2015).

The electric vehicles industry is at a nascent stage in India. EVs currently account for less than 1% of total vehicle sales and have potential to grow to more than 5% in a few years. At present, there are more than 4,00,000 electric two wheelers and a few thousand electric cars on Indian roads ("High Vat", 2014). The industry volumes have been fluctuating, mostly depending on the incentives offered by the government. Many serious players (Hero Eco, Mahindra Reva, Electrotherm, Avon, Lohia, Ampere etc.) are continuing with the mission and trying to enforce positive change.

More than 95% of electric vehicles on Indian roads are low-speed electric scooters (less than 25km/hr) that do not require registration and licenses (Bhakta, 2015). Almost all electric scooters run on lead batteries to keep the prices low. Besides government subsidies, battery failures and low life of batteries have become major limiting factors for uptake of low-speed scooters. Mahindra Reva is the only electric car manufacturer in India, and their new model E2O runs on lithium batteries. Manufacturers like Mahindra Reva and Hero have taken initiatives to install charging stations (Mahindra Reva Electric Vehicles Pvt. Ltd., 2013), but at limited places. Players like Lohia and Electrotherm have developed electric three wheelers. Ampere and Hero have entered the electric cycles segment.

1.11.1 National Electric Mobility Mission Plan

The Indian government has recognized the importance of electric vehicles and launched a program in 2013 named "National Electric Mobility Mission Plan (NEMMP) 2020". It aims to achieve national fuel security by promoting electric vehicles (comprising electric two-wheelers, electric three-wheelers, battery vans, hybrid vehicles and electric bicycles) and infrastructure to support EV growth as a solution to the growing problem of urban pollution ("National Electric", 2015).

The Indian government recognizes the potential for the electric vehicle to address transport issues and has a desire to implement the technology to ease congestion in urban areas. Some of the targets set for reducing carbon emissions are:

- Reduction of the use of liquid fuel by 7 to 8 million tons by 2020
- Reduce vehicle emissions (CO_2) by 1.5% by 2020

The government plan has set out targets to be met by 2020 as below:

- India to be a major global player by 2020 in the electric vehicle area,

- Provide employment to 25 million people by 2016 in the automotive sector,
- Production of 6 to 7 million electric vehicles (incl. all types) on the road by 2020,
- Expected 4 to 5 million are to be two-wheeled electric vehicles,
- To reduce dependence on oil imports for the transport sector,
- To create automotive "Centre of Excellence" hubs to promote EVs throughout India.

To support its strategic plan, the Indian government is keen to develop a collaborative effort between the government, domestic automotive industry and R&D agencies.

The industry is almost ready for take-off with the support of incentives. It is expected that NEMMP will be implemented within a year and with that, the industry may witness a quantum leap in volumes and technology. The Society of Manufacturers of Electric Vehicles (SMEV) sees a great opportunity with EVs in reducing the carbon footprint, dependence on crude oil imports, creating jobs and building a new technology knowledge hub in India.

The Indian government also launched the Faster Adoption and Manufacturing of Hybrid and Electric Vehicles (FAME) Scheme to provide incentives for two wheelers and four wheelers. This scheme is part of NEMMP and is expected to provide support of INR 795 crore for the financial year of 2015-16 and 2016-17 (First post, 2015). This Scheme was launched to have faster adoption, domestic technology development and help in manufacturing cleaner electric vehicles including:

- Full/mild hybrid (HEVs)
- Plug-in hybrid (PHEVs)
- Pure Electric (BEVs)

1.11.1.1 Purpose of NEMMP 2020

- Today, 87% of India's fossil fuel is imported with economic growth dependency on imports, which are likely to increase by 92% by 2020.
- With the sharp rise in the number of vehicles and as per the projections of International Energy Agency (IEA), it is estimated that 3/4th of the projected future increase in oil demand will be from transportation.

- Hybrid and electric vehicles are expected to have 35% to 45% lower emissions in comparison with conventional IC engines.
- Based on the studies, the government has found out that by 2020, demand for vehicles will increase by 41 million (32 million two wheelers and 9 million four wheelers) from the current level of 13 million and 3 million units respectively.

1.11.1.2 NEMMP 2020 – Approach

Types of incentives planned under the scheme:

- **Demand-side incentives:** focus on creating demand byincentivizing consumers.
- **Supply side incentives:** focus on creating a supply of affordable vehicles into the market.
- **R & D:** focus on creating technology capability to achieve localization and domestic manufacturers.
- **Charging Infrastructure:** focus on creating a conducive environment for mass adoption of electric vehicles.
- **Pilot projects:** focus on creating awareness, kick-starting adoption, creating test market conditions, developing viable business models and obtaining product feedback from customers.

1.12 Electric vehicle models available in India

The electric automobile industry in India is extremely small. Mahindra REVA is the only manufacturer of electric vehicles in India in the four wheeler category. In the two wheeler market, the manufacturers are Lohia Auto, Hero Electric, YoBykes, and Ampere (Top 5, 2013). The following section lists the top five electric/hybrid cars models in India (Naik, 2015):

1. **Mahindra e20:** Mahindra launched its first electric model named e20 in 2013 in the Indian market. The product was manufactured at the Bengaluru plant. The price range for Mahindra e20 is INR 4.79 lakh – INR 5.34 lakh.

2. **Toyota Prius:** Toyota launched its first hybrid car in 2012 in India. Petrol and electricity are the two energy sources for its engine. The price of this model in the Indian market is very high because of import duty and the price range is between INR 38.10 lakh - INR 39.80 lakh

3. **Toyota Camry Hybrid:** Camry Hybrid delivers a mileage of 19.6 Km./l (ARAI figure) – claimed by Toyota with a price of INR 31.19 lakh

4. **BMW i8:** In August 2014, BMW started deliveries of its i8 model in the Indian market and sold 1,741 units till December, 2014. The cost of this product is INR 2.29 crore.

5. **Maruti Suzuki Ciaz Hybrid:** Maruti Suzuki Ciaz launched its Smart Hybrid vehicle in India in September 2015 with the price range of INR 8.23 lakh – INR 10.17 lakh.

Table 1: Car models available in the world

Model	Top Speed (km/h)	Market release (Year)
NICE Mega City	64	2006
Mahindra e20	82	2013
Stevens Zecar	90	2008
BMW Brilliance Zinoro 1E	130	2014
BolloréBluecar	130	2011
Ford Focus Electric	135	2011
Lightning GT	200	2013
Nissan Leaf	150	2010
Renault Fluence ZE	135	2010
Renault Zoe	135	2012
Smart electric drive	125	2009
Tesla Model SP85 kW·h	214	2012
VenturiFétish	200	2006
Volkswagen e-Golf	145	2014
Volkswagen e-Up!	130	2013
Detroit Electric SP:01	249	2014

Sourc – websites of above mentioned companies

1.13 Chapterisation of the study

Chapter 1: 'Introduction' provides the background of electric vehicles (EV), important stakeholders, various risks associated with EV, global overview of EV including the Indian government's plans and various EV models available across the world.

Chapter 2: 'Consumer Survey' focuses on participants' demographics (age, gender, education, monthly income, region and daily travel distance) and the various factors which affect the purchasing decisions for vehicles. It also highlights the role of government in the promotion of electric vehicles and consumer willingness to pay a premium to purchase an electric vehicle in the future.

Chapter 3: 'Consumer Perceptions' explains the role of region (geography), cities, gender, education, income, age group and environment in the selection of vehicle. It also highlights the respondent's perception towards environment related issues along with the list of key influencing factors.

Chapter 4: 'Consumer Survey Outcomes and Recommendations' highlights the findings and conclusion of the study. General and institutional recommendations are also discussed in this chapter.

Chapter 5: 'Consumer Forums' highlights people's perception of various aspects of electric vehicles. The responses were collected from Lucknow and New Delhi.

Chapter 2

Consumer Survey

Fossil fuels comprise of three types of energy resources – coal, oil/petroleum and natural gas and have become a major energy source across the world. Over consumption of fossil fuel by transport and industrial sector has raised many serious issues, especially for the environment, and affected the ecosystem of human beings. Fossil fuel disadvantages include an adverse impact on human health which has resulted in severe diseases like asthma, lung cancer, chronic obstructive pulmonary disorder or COPD etc. (Conserve-Energy-Future). It becomes extremely important to understand the perception of human beings who are part of the transport sector either by owning any vehicle or by using public transport.

2.1 Study objectives

This study was conducted to understand the awareness level among young professionals in India related to electric vehicles (EVs) and the reasons for making a decision to buy an EV. The results can be useful for policy makers, car manufacturers and researchers on the fundamental reasons controlling the use of EV so that steps can be taken to improve usage. The major objectives of the study are:

- Identification of factors driving the use of electric vehicles.
- Feasibility of using electric cars in various cities of India.
- To determine the target group based on demographic characteristics for promoting electric vehicles.
- Availability of electric vehicle related to purchase price, fuel consumption and operational cost of such vehicles.
- Identification of factors such as personal preferences, range anxiety, children's influence on environmentally friendly transport.

2.2 Methodology

The study was exploratory and analytical in nature. Since the actual target population number was not known, the authors used convenience sampling. In total, 2281 respondents from across India took part in this study. The target audience for the study was middle-class and upper-middle-class populations because they might become future buyers. Primary data was collected during Nov 2013 to Apr 2015.

The descriptive analysis was done by using frequency distribution and cross tabulation in the Statistical Package for the Social Sciences (SPSS version 22) software. Frequency tables and cross tabulation were constructed to display results with respect to research objectives.

2.3 Designing of questionnaire

Before designing the questionnaire, more than 200 journals, books, reports and articles were reviewed. A detailed questionnaire was developed with the consultation of important stakeholders. The detailed questionnaire is attached as Annexure 4. The objective of this survey was to assess the general public's knowledge and opinion of EVs by gathering information about the following:

- General awareness of the current leading EV technologies.
- Opinions on the societal impacts of EVs.
- Opinions on how different factors influence consumer decisions for vehicles.

The questionnaire was broadly divided into three parts as described in the below section-

In **Part 1** of the questionnaire, questions related to socio-demographic characteristics were asked. Questions related to gender, age group, education level, their current occupation, the number of family members and their current vehicle ownership were asked.

In **Part 2** of the questionnaire, we asked questions to understand their driving behavior along with the important sources of information and critical factors which control their decision while purchasing any vehicle. The questions were, how much do they drive per day, how do they become aware of new technology developments, who influences their vehicle purchase decision, when a new vehicle becomes available for purchase, what do they do, importance of critical factors which are considered for choosing the kind of vehicle they drive, perception about vehicular pollution.

In **Part 3** of the questionnaire, questions were asked to understand the perception of respondents related to electric vehicles. The questions were, how interested would they be in purchasing an electric motor vehicle once they become easily available in the next couple of years, do they think transport policy of the State and Central Government should encourage electric vehicles, identification of important variables that can be considered for purchasing or leasing an electric vehicle in the future. We also asked questions related to mapping the perception of respondents about the environmental impact of motor vehicles and their willingness to reduce it by adopting such vehicles in the near future. Respondents were also asked if they would be keen to pay a premium to purchase a electric vehicle, if yes, then what will be the percentage.

2.4 Consumer demographics

2.4.1 Age group of respondents

The data was collected from 2281 respondents from 16 cities and 9 states of India. All the data was collected through convenience sampling. As described in figure 1, this resulted in a sample where 55.5% of respondents were 18 to 30 years old, 36.4 % of respondents were 31 to 60 years old and 6.2% of the respondents were more than 61 years old. However, 1.9% of the respondents didn't disclose their age group.

Figure 1: Age group of respondents

Age group of the respondents

2%
6%
36%
56%

18-30
31-60
61 and Above
Not Disclosed

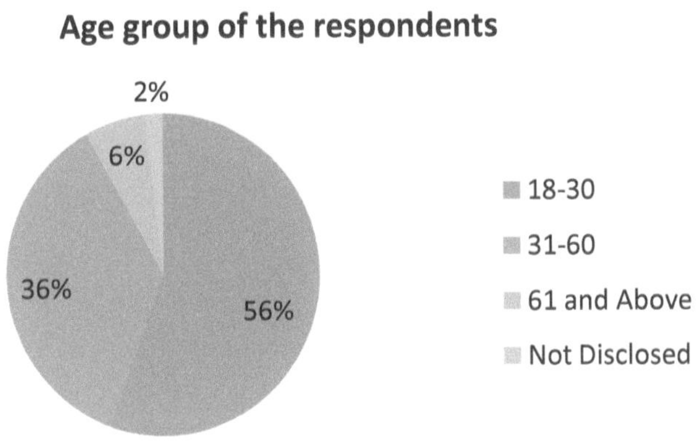

2.4.2 Gender

We have taken responses from 2281 respondents by involving 55.8% male and 42.52% female. However, 1.66% didn't disclose their gender. (Figure 2)

Figure 2: Gender of the respondents

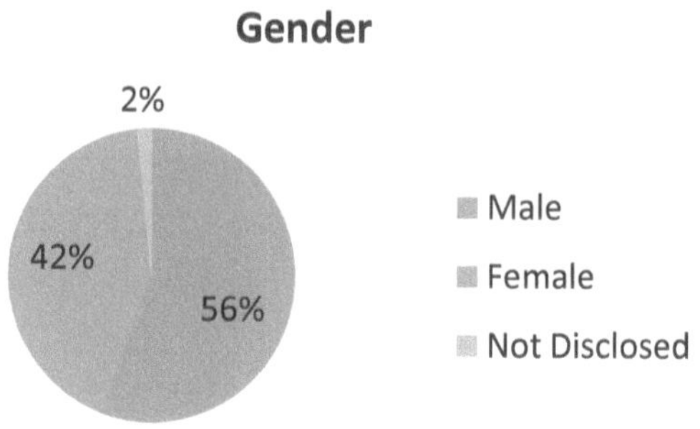

2.4.3 Education level and background of respondents

Out of 2281 respondents, 1056 (46.2%) had a bachelor's degree, 881 (38.62%) had a master's degree, 311 (13.63%) were students (Figure 3). Data indicates that the respondents were well educated, highly sophisticated, focused in their opinion and well informed people. They are decision makers, opinion builders, and in many cases role models in their surroundings and social circles.

Figure 3: Education level of respondents

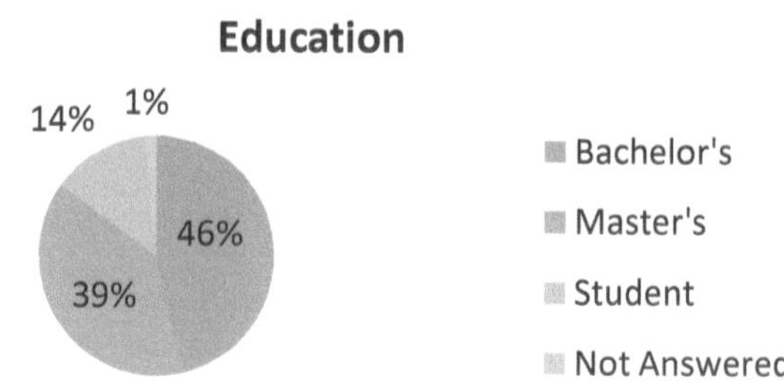

2.4.4 Monthly income

In the survey, we found that out of 2281 respondents, 1216 (53.3%) shared that they have a monthly income between INR 25,000 – 50,000, 702 (30.8%)

shared that their income is between INR 50,000 – 100,000, 137 (6%) shared that they have income more than INR 1 lakh (Figure 4).

Figure 4: Monthly income of the respondents

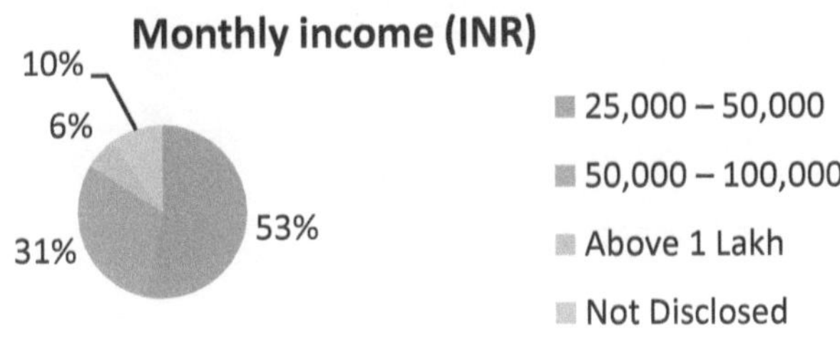

2.4.5 Residence (Region)

The data was collected from 9 states as mentioned in figure 5 and from 16 districts as mentioned in figure 6. In total 2281 respondents were covered and their preferences on electric vehicle were recorded and analyzed.

Figure 5: States covered during the study

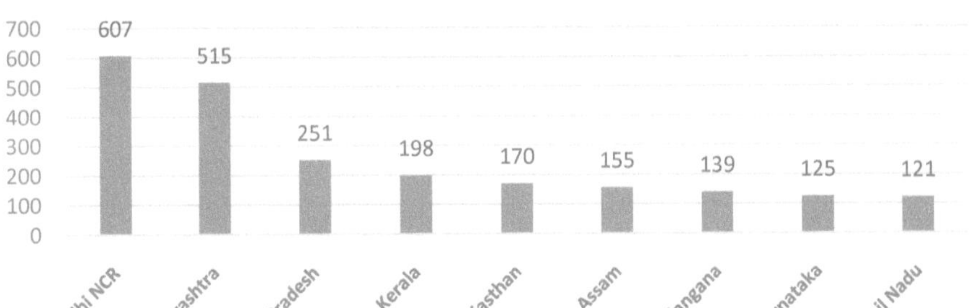

Figure 6: Cities covered during the study

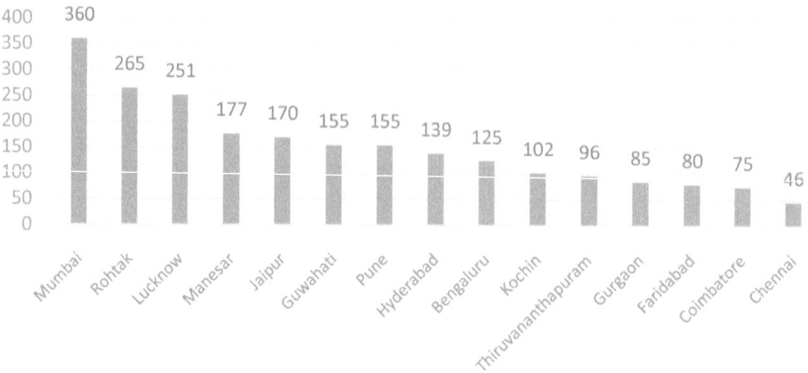

2.4.6 Estimated average daily drive

Out of 2281 respondents, 348 (15.3%) respondents shared that they travel up to 10 km in a day, 676 (29.6%) shared that they travel between 11-20 km in a day, 428 (18.8%) shared that they travel between 21-30 km in day, 489 (21.4%) shared that they travel between 31-60 km in a day, 200 (8.8%) responded that they travel more than 60 km in a day. It can be easily interpreted that the percentage of respondents who travels less than 60 km per day is 85.1% (Figure 7). It is clear that this group if motivated properly and given the sufficient services, can think of using electric vehicles.

Figure 7: Estimated average daily drive of the respondents

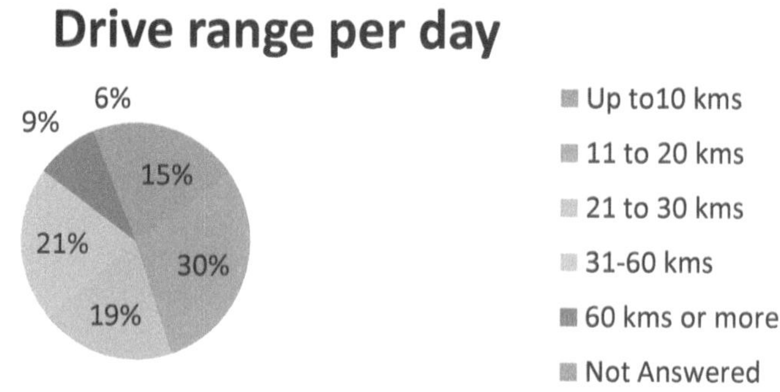

2.5 Consumer perception about vehicles

2.5.1 Awareness about new technology

We asked this question to understand the information sources of respondents. We tried to give them most of the information sources available in the market as options. 1319 respondents shared that their information source is television, 1339 respondents shared that they get information from newspapers, 381 respondents get awareness from radio, 801 get information from various magazines, 1405 one get information from the internet, 241 get information from word of mouth marketing, 197 shared that they get information from other sources like their friends, relatives, family members and various hoardings (Figure 8).

Figure 8: Awareness source about new technologies

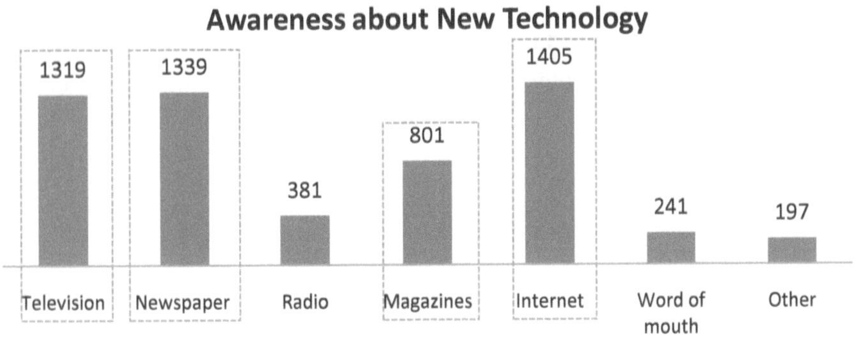

Awareness about New Technology

1319
1339
381
801
1405
241
197

Television
Newspaper
Radio
Magazines
Internet
Word of mouth
Other

2.5.2 Influence on purchasing decision

Out of 2281 respondents, 1162 respondents (50.9%) consult with elders while taking the decision about purchasing any vehicle. 287 respondents (12.6%) go with the joint decision of their families. 769 respondents (33.7%) shared that they discuss the decision with their children while purchasing any vehicle (Figure 9). It can be concluded from the above data that every member of the family has a say indecisions related to purchases, including those on vehicles.

Figure 9: Vehicle purchase decision

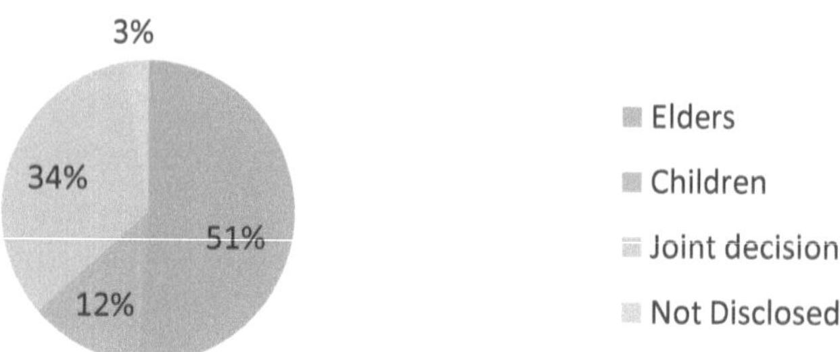

2.5.3 About new technologies

Out of 2281 respondents, 990 (43.4%) shared that they will read the reviews and buy if the reviews are favorable, 852 (37.4%) shared that they prefer to wait till the new technology is widely accepted and proven. 321 (14.1%) shared that they will be among the first ones to purchase the new vehicle (Figure 10).

Figure 10: Views about purchasing a new technology vehicle

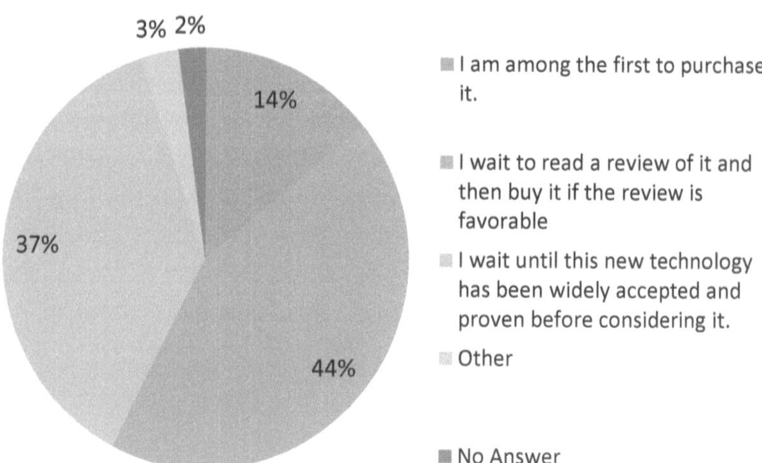

2.5.4 Important considerations (for choosing the vehicles)

The respondents disclosed their concerns for fuel efficiency, safety, vehicle power and purchase price. Reliability, vehicle size, expected operating costs, reputation of a particular vehicle, fuel type, vehicle emission and pollution were also factors that are important to them.

Figure 11: Important factors for choosing vehicles

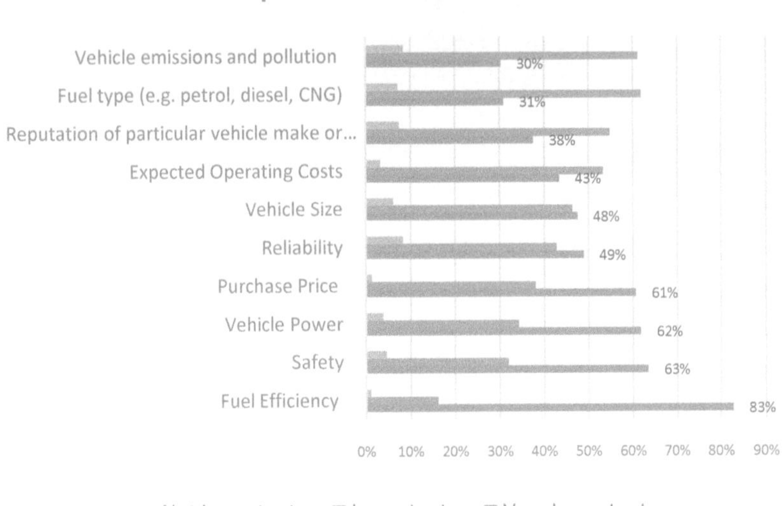

2.5.5 Current ownership of electric vehicle

Only 3.7% respondents had an electric vehicle at home, 92.8% do not use an electric vehicle as of now (Figure 12). Respondents said that the price, availability and reliability are stopping them from purchasing an electric vehicle. Traffic is also an issue in metro cities which stops them from going for an electric vehicle.

Figure 12: Current ownership of electric vehicle

2.6 Cost of purchasing EVs is not the only key factor today

2.6.1 Availability of electric vehicles in the market

Most of the respondents shared that they don't have many options in the four wheeler segment except Mahindra – Reva, but for the two-wheeler, they know of some options like Yo Bikes, Hero Electric etc. They also highlighted that e-rickshaws are becoming popular in metro cities and tier-1 cities.

2.6.2 Interest in new electric vehicles

Out of 2281 respondents, 650 (28.5%) shared that they will be very interested in purchasing an electric motor vehicle once they become easily available in next couple of years, 914 (40.1%) shared that they will be somewhat interested, 287 (12.6%) shared that they will not be very interested in purchasing of such a vehicle, 207 (9.1%) said that they are not planning to buy any vehicle in near future (Figure 13).

Figure 13: Interest in purchasing electric vehicle

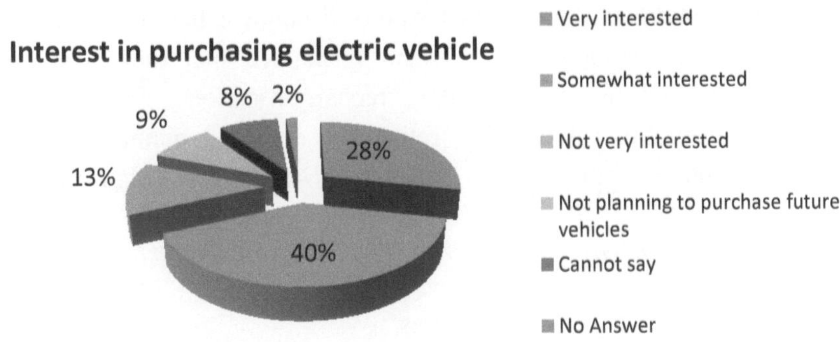

2.6.3 Reasons for purchasing electric vehicle in the future – Environmental factors

Out of 2281 respondents, only 217 (9.7%) were not interested in "driving a vehicle with more advanced or innovative technology". 696 (31%) said that "it is very important to reduce the dependence on gasoline", 1004 (44.7%) said that it is important. 618 (27.5%) said that it is very important to reduce the impact on the environment by introducing electric vehicles in the market, 1442 (64.2%) said that it is important. 362 (16.1%) respondents said that "saving money on the cost of operation is very important for them", 986 (43.9%) said that it is important (Figure 14).

Figure 14: Reasons for purchasing an electric vehicle in future – Environmental factors

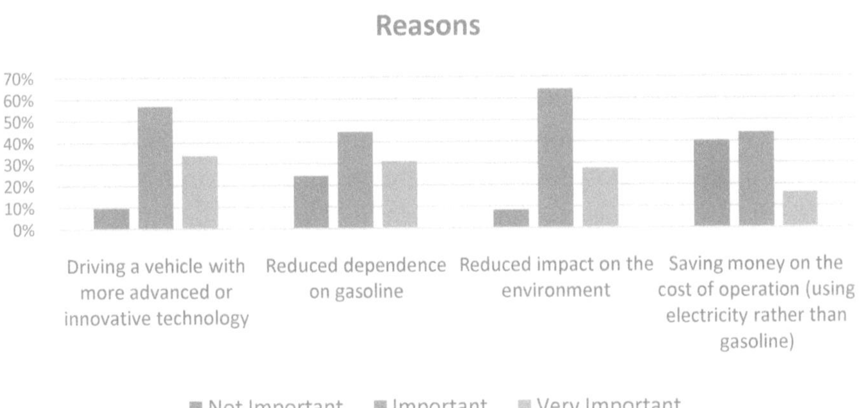

2.6.4 Concerns about electric vehicles – Vehicle attributes

When we asked about the factors that would discourage the customer from buying an electric vehicle, most of them responded that they would be concerned about various factors like "recharge stations, higher price, availability of desirable vehicle size, reliability, on-going maintenance and operation cost, and the ability to carry heavy loads sometimes" (Figure 15).

Figure 15: Conerns about electric vehicles – Vehicle attributes

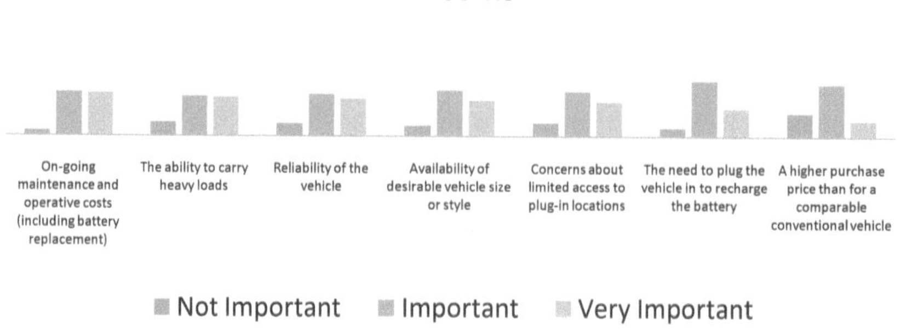

2.6.5 Views about environmental impact

61.8% disagreed with the statement that "cars, minivans, pickups and SUVs are an important source of air pollution" whereas 31.9% agreed with this statement. 51.1% agreed with the statement that "cars, minivans, vans, pickups and SUVs are an important source of the greenhouse gases that many scientists believe are warming the earth's climate," whereas 29.2% differed with the statement. Similarly, 45.8% agreed that "exhaust from cars, minivans, vans, pickups and SUVs is an important source of the pollution that causes asthma and makes asthma attacks worse" and 30.8% disagreed. 42.2% agreed that "government rules allow minivans, vans, pickups and SUVs to pollute more than passenger cars, for every gallon of gas used" while 41.4% disagree with it. Finally, 34.7% agreed that "government rules require minivans, vans, pickups and SUVs to meet the same miles-per-gallon standards as passenger cars" while 40.1% disagree with it (Figure 16).

Figure 16: Views about environmental impact

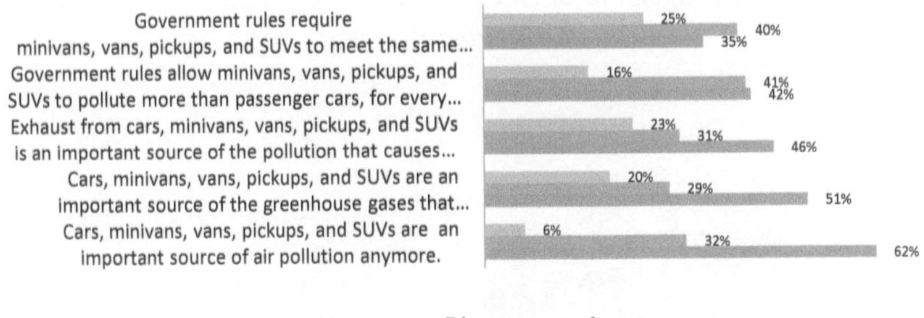

2.6.6 Premium for electric vehicle

We observed that people have started thinking about the environment and climate change. They want to contribute to the solution. We asked respondents if they would pay a premium for new technology or electric vehicles. Out of 2281respondents, 637 (27.9%) are willing to pay a premium of 10%, 931 (40.8%) can pay 15% premium, 598 (26.2%) can pay 20% premium. However, 115 (5%) refused to pay any premium for this vehicle (Figure 17).

Figure 17: Willing to pay apremium for an electric vehicle

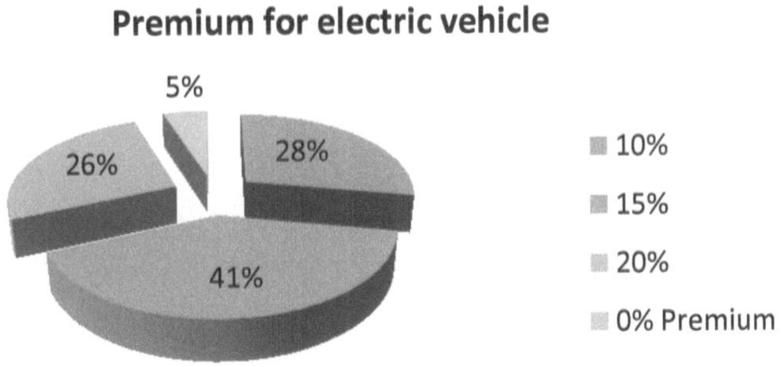

Premium for electric vehicle

5%
28%
26%
41%

10%
15%
20%
0% Premium

2.6.7 Role of State and Central Government

54.4% of the total respondents were in favor of promoting an EV-friendly transport policy both for public and private transport. 30.5% respondents were in favor of only public transport. 13% of the respondents did not find the current policies sufficient to promote the electric vehicle in the market (Figure 18).

Figure 18: Encouragement of electric vehicles by State and Central Government

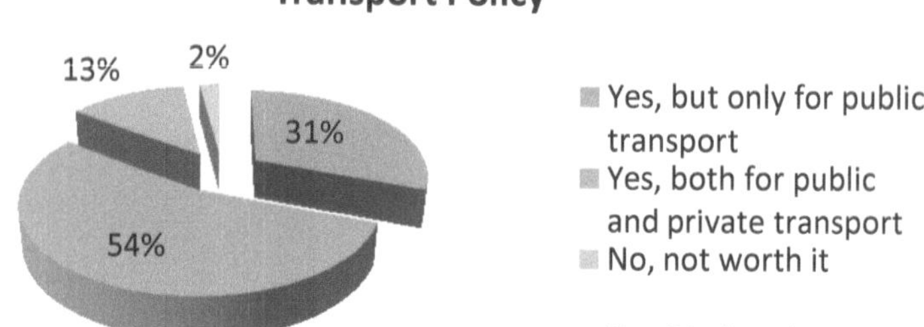

Consumer Perceptions

To understand the impact of consumer perceptions or preferences on the selection of electric vehicle, we have analyzed the responses by region, city, age, education, income groups and gender. We also analyzed the variables to understand the role of environment in purchasing of electric vehicles.

3.1 Regional perceptions

To understand the regional perceptions, we asked multiple questions to the respondents. The questions were asked to know the average daily travel, different types of information sources, who influences their purchasing decisions, their views on purchasing electric vehicles, what they think about the role of State and Central Government, important factors for purchasing any vehicle, different types of transport issues, various reasons that can influence their decisions to purchase electric vehicles in the future, impact of transport sector on the environment, especially vehicular pollution and finally the perception on the premium for electric vehicles. The results were analyzed region wise and summarized in the below section.

3.1.1 Estimated average daily drive

Table 2 highlights thatin South India, a majority of the people travel upto 30 km in a day. In East India, people travel shorter distances to complete their routine work: almost 75% of the respondents shared that they travel upto 20 km in a day to complete their work. In West India and North India, the situation is different from the other two regions. In these two regions, people travel more to complete their work. Data revealed that approximately 42% respondents in the West region and 37% respondents in the North region travel more than 30 km in a day for their routine work.

Table 2: Estimated average daily travel – Zone-wise

		Zone							
		South India		West India		North India		East India	
		Count	Column N %	Count	Column N %	Count	Column N %	Count	Column N %
Drive per day	Up to 10 km	135	30.2%	32	6.2%	109	10.9%	72	41.1%
	11 to 20 km	134	30.0%	143	27.8%	335	33.4%	64	36.6%
	21 to 30 km	100	22.4%	122	23.7%	190	18.9%	16	9.1%
	31-60 km	59	13.2%	147	28.5%	262	26.1%	21	12.0%
	60 km or more	19	4.3%	71	13.8%	108	10.8%	2	1.1%
Cells highlighted represent the highest percentage across zones									

3.1.2 Information Source

In Table 3, the data revealed that there is a significant difference which exists between sources of information among regions. In the South region, the internet is the main source of information followed by television and newspapers. Similarly in the North region respondents shared that the major information source is the internet followed by newspapers, television, magazines and radio. In the West region, television and internet are the main information sources. The region has other important information sources as well such as newspapers, magazines and radio. In East India, the major source of information is television followed by newspapers, the internet, magazines, word of mouth and radio.

Table 3: Information source for new technology

Various sources of information	Zone							
	South India		West India		North India		East India	
	Count	Column N %	Count	Column N %	Count	Column N %	Count	Column N %
Television	298	54.9%	309	60.0%	585	57.1%	127	70.9%
Newspapers	265	48.8%	277	53.8%	691	67.5%	106	59.2%
Radio	23	4.2%	103	20.0%	225	22.0%	30	16.8%
Magazines	154	28.4%	235	45.6%	353	34.5%	59	33.0%
Internet	332	61.3%	304	59.0%	698	68.2%	71	39.7%
Word of mouth	100	18.4%	29	5.6%	71	6.9%	41	22.9%
Other	46	8.5%	63	12.2%	82	8.0%	6	3.4%
Cells highlighted represent the highest percentage across zones								

3.1.3 Influence on purchasing decision

Data in Table 4 shows that in the South region people take joint decisions while purchasing any new vehicle, however in the other three regions, they consult with elders while taking a decision about purchasing any vehicle. It can be concluded from the above data that every member of the family has a say in the decision-making process while purchasing vehicles in all regions.

Table 4: Vehicle purchase decision – influence source

		Zone							
		South India		West India		North India		East India	
		Count	Column N %	Count	Column N %	Count	Column N %	Count	Column N %
Who influences your vehicle purchase decision	Elders	231	45.4%	285	55.3%	516	50.7%	130	73.4%
	Children	25	4.9%	73	14.2%	176	17.3%	13	7.3%
	Joint decision	253	49.7%	157	30.5%	325	32.0%	34	19.2%

Cells highlighted represent the highest percentage across zones

3.1.4 About new vehicles

Respondents from the East region shared that they were among the first to purchase a new vehicle. Respondents from the South region and West region shared that they will wait to read the reviews and then will buy it if the reviews are favorable. However the respondents from the North region shared that they will wait until the new technology is widely accepted and proven before purchasing a new vehicle (Table 5).

Table 5: When a new vehicle becomes available for purchase, what do you do?

		Zone							
		South India		West India		North India		East India	
		Count	Column N %	Count	Column N %	Count	Column N %	Count	Column N %
When a new vehicle becomes available for purchase, what do you do?	I am among the first to purchase it.	48	8.9%	54	10.6%	143	14.2%	76	42.9%
	I wait to read a review of it and then buy it if the review is favorable	248	45.8%	297	58.2%	390	38.8%	55	31.1%
	I wait until this new technology has been widely accepted and proven before considering it.	224	41.4%	151	29.6%	433	43.1%	44	24.9%
	Other.	21	3.9%	8	1.6%	38	3.8%	2	1.1%
Cells highlighted represent the highest percentage across zones									

3.1.5 Interested in new electric vehicle

Table 6 revealed that respondents from the East region are very interested in purchasing an electric motor vehicle once they become easily available in the next couple of years while respondents from other three regions were skeptical about purchasing such electric vehicles. They were "somewhat interested" in purchasing such vehicles.

Table 6: Interest level in purchasing of electric vehicle

		Zone							
		South India		West India		North India		East India	
		Count	Column N %	Count	Column N %	Count	Column N %	Count	Column N %
How interested would you be in purchasing an electric motor vehicle once they become easily available in the next couple of years?	Very interested	188	34.9%	126	24.9%	227	22.3%	109	60.6%
	Somewhat interested	233	43.2%	175	34.5%	457	44.8%	49	27.2%
	Not very interested	44	8.2%	90	17.8%	144	14.1%	9	5.0%
	Not planning to purchase future vehicles	27	5.0%	60	11.8%	110	10.8%	10	5.6%
	Cannot say	47	8.7%	56	11.0%	82	8.0%	3	1.7%
Cells highlighted represent the highest percentage across zones									

3.1.6 Role of State and Central Government

Table 7 highlights that respondents from the East region shared that the State and Central Government should encourage electric vehicles only for public transport; however respondents from other three regions shared that the State and Central Government should encourage electric vehicles for both public and private transport by focusing on transport policy. Approximately 15.4% respondents from the North region and 14.8 % respondents from West region did not find the current policies sufficient to promote electric vehicles in the market.

Table 7: Encouragement of electric vehicles by State and Central Government

		Zone							
		South India		West India		North India		East India	
		Count	Column N %	Count	Column N %	Count	Column N %	Count	Column N %
Do you think transport policy of the State and Central Government should encourage electric vehicles?	Yes, but only for public transport	121	22.7%	129	25.5%	324	32.0%	121	67.6%
	Yes, both for public and private transport	370	69.4%	302	59.7%	534	52.7%	35	19.6%
	No, not worth it	42	7.9%	75	14.8%	156	15.4%	23	12.8%
Cells highlighted represent the highest percentage across zones									

3.1.7 Factors important for purchase of personal vehicles - Region wise

Table 8 shows that factors such as fuel efficiency, reliability and vehicle power were very important for purchasing of a personal vehicle in all regions. For South, North and West region, safety is a very important factor, however for East region respondents, it is an important factor. Vehicle size along with the market reputation of a particular vehicle is an important factor which influences the purchasing habits of respondents.

Table 8: Important factors for purchasing of personal vehicles by region

Variables	Perceptions	Zone							
		South India		West India		North India		East India	
		Count	%	Count	%	Count	%	Count	%
Fuel efficiency	Very Important	469	85.0%	413	81.0%	829	80.7%	165	91.7%
	Important	80	14.5%	85	16.7%	187	18.2%	15	8.3%
	Not Important	3	.5%	12	2.4%	11	1.1%	0	0.0%
Safety	Very Important	475	86.2%	320	62.4%	546	53.2%	35	19.4%
	Important	75	13.6%	179	34.9%	466	45.4%	144	80.0%
	Not Important	1	.2%	14	2.7%	14	1.4%	1	.6%
Vehicle power	Very Important	294	53.7%	318	62.0%	673	65.6%	152	84.4%
	Important	240	43.9%	154	30.0%	304	29.6%	26	14.4%
	Not Important	13	2.4%	41	8.0%	49	4.8%	2	1.1%
Purchase price	Very Important	264	48.0%	272	53.0%	422	41.2%	25	13.9%
	Important	274	49.8%	215	41.9%	567	55.4%	153	85.0%
	Not Important	12	2.2%	26	5.1%	35	3.4%	2	1.1%
Reliability	Very Important	367	67.6%	260	50.7%	612	59.9%	156	86.7%
	Important	166	30.6%	221	43.1%	363	35.5%	24	13.3%
	Not Important	10	1.8%	32	6.2%	47	4.6%	0	0.0%
Vehicle size (to accommodate passengers or cargo)	Very Important	152	27.9%	196	38.2%	324	31.7%	26	14.4%
	Important	336	61.7%	271	52.8%	640	62.6%	152	84.4%
	Not Important	57	10.5%	46	9.0%	58	5.7%	2	1.1%
Expected operating costs	Very Important	222	40.5%	169	33.1%	539	52.6%	145	80.6%
	Important	292	53.3%	286	56.0%	437	42.7%	34	18.9%
	Not Important	34	6.2%	56	11.0%	48	4.7%	1	.6%
Reputation of particular vehicle make or model	Very Important	162	30.5%	164	32.0%	329	32.2%	27	15.0%
	Important	300	56.5%	291	56.7%	631	61.8%	151	83.9%
	Not Important	69	13.0%	58	11.3%	61	6.0%	2	1.1%
Fuel type (e.g. petrol, diesel, CNG)	Very Important	230	42.1%	178	34.8%	551	53.8%	147	81.7%
	Important	257	47.1%	278	54.3%	400	39.1%	32	17.8%
	Not Important	59	10.8%	56	10.9%	73	7.1%	1	.6%
Vehicle emissions and pollution	Very Important	302	55.1%	172	33.5%	341	33.3%	36	20.0%
	Important	200	36.5%	279	54.4%	621	60.7%	143	79.4%
	Not Important	46	8.4%	62	12.1%	61	6.0%	1	.6%

Cells highlighted represent the highest percentage across zones

3.1.8 Transport issues

For better understanding the issues related to the transportation system, questions focusing on the problems faced by the respondents were asked. Various variables were included such as "congestion and noise; vehicle emissions and pollution; traffic speed; global warming; fuel import" etc. These variables highlighted the preferences of respondents while buying any vehicle. In Table 9, we have explained the various issues based on regional perceptions.

Table 9: Transport issues based on regions

Variables	Perceptions	Zone							
		South India		West India		North India		East India	
		Count	%	Count	%	Count	%	Count	%
Traffic congestion that you experience while driving	Not a Problem	55	10.2%	147	28.7%	347	33.8%	124	68.9%
	Problem	231	42.8%	301	58.8%	566	55.2%	40	22.2%
	Major Problem	254	47.0%	64	12.5%	113	11.0%	16	8.9%
Traffic noise that you hear at home, work or school	Not a Problem	53	9.7%	54	10.5%	111	10.8%	5	2.8%
	Problem	314	57.5%	337	65.8%	641	62.5%	147	81.7%
	Major Problem	179	32.8%	121	23.6%	274	26.7%	28	15.6%
Vehicle emissions that affect local air quality	Not a Problem	24	4.4%	61	11.9%	214	20.9%	119	66.1%
	Problem	212	39.1%	260	50.8%	523	51.1%	36	20.0%
	Major Problem	306	56.5%	191	37.3%	286	28.0%	25	13.9%
Vehicle emissions that contribute to climate change	Not a Problem	30	5.5%	73	14.3%	112	11.0%	5	2.8%
	Problem	214	39.4%	243	47.5%	550	53.8%	143	79.4%
	Major Problem	299	55.1%	196	38.3%	360	35.2%	32	17.8%
Unsafe communities because of speeding traffic	Not a Problem	31	5.8%	60	11.7%	201	19.6%	113	62.8%
	Problem	261	48.9%	273	53.3%	521	50.9%	40	22.2%
	Major Problem	242	45.3%	179	35.0%	302	29.5%	27	15.0%
Importing much of our oil from foreign countries	Not a Problem	59	11.0%	76	14.8%	92	9.0%	9	5.0%
	Problem	208	38.7%	206	40.2%	513	50.1%	142	78.9%
	Major Problem	270	50.3%	230	44.9%	418	40.9%	29	16.1%

Cells highlighted represent the highest percentage across zones

3.1.9 Various reasons for purchasing electric vehicles in future

Table 10 highlights that "saving money on the cost of operation (using electricity rather than gasoline)" is not important for the North and East region, however, it is important for the South and West regions. Respondents from all the regions were concerned about the environmental impact and highlighted that electric vehicle will reduce pollution in the near future. Respondents from West, North and East India felt that "it is important to reduce the impact of transport on the environment", however, respondents from South India felt that "it is very important to reduce the impact of transport on the environment".

Respondents from South India and East India feel that "it is important to reduce dependence on gasoline". Respondents from North India and East India felt that "it is important to drive a vehicle with more advanced or innovative technology".

Table 10: Regional preferences for various attributes

Variables	Perceptions	Zone							
		South India		West India		North India		East India	
		Count	%	Count	%	Count	%	Count	%
Saving money on the cost of operation (using electricity rather than gasoline)	Not Important	53	9.9%	175	34.5%	520	50.9%	149	82.8%
	Important	281	52.3%	276	54.4%	407	39.9%	22	12.2%
	Very Important	203	37.8%	56	11.0%	94	9.2%	9	5.0%
Reduced impact on the environment	Not Important	34	6.4%	32	6.3%	108	10.6%	11	6.1%
	Important	217	40.6%	345	68.0%	729	71.3%	151	83.9%
	Very Important	284	53.1%	130	25.6%	186	18.2%	18	10.0%
Reduced dependence on gasoline	Not Important	36	6.7%	42	8.3%	333	32.6%	134	74.4%
	Important	301	56.3%	243	47.9%	431	42.1%	29	16.1%
	Very Important	198	37.0%	222	43.8%	259	25.3%	17	9.4%
Driving a vehicle with more advanced or innovative technology	Not Important	58	10.8%	59	11.7%	84	8.2%	16	8.9%
	Important	267	49.9%	222	43.9%	632	61.8%	150	83.3%
	Very Important	210	39.3%	225	44.5%	307	30.0%	14	7.8%

Cells highlighted represent the highest percentage across zones

As explained in Table 11, the common issues that came out were "recharge stations, higher price, availability of desirable vehicle size, reliability, on-going maintenance, operating costs and the ability to carry occasionally heavy loads" which are restricting the demand for EVs in the Indian market. For East India respondents, it is not important to purchase an EV by paying a higher price than for a comparable conventional vehicle, for the rest of the regions, i.e. South, West and North, it is important to pay a higher price for an EV than for a comparable conventional vehicle.

For South India respondents, reliability of the vehicle is important, however, for the Northern and Eastern region, vehicle size is important. Respondents from West India were more concerned about limited access to plug-in locations and considered this an important factor. Maintenance and the operating cost were a very important factor for the West and an important factor for the North and East region. However, none of the regions were concerned about the heavy load capacity of EVs.

Table 11: Factors supporting purchase of electric vehicles by region

Variables	Perceptions	Zone							
		South India		West India		North India		East India	
		Count	%	Count	%	Count	%	Count	%
A higher purchase price than for a comparable conventional vehicle	Not Important	75	14.3%	110	21.7%	297	29.1%	90	50.0%
	Important	308	58.9%	306	60.4%	580	56.9%	71	39.4%
	Very Important	140	26.8%	91	17.9%	143	14.0%	19	10.6%
The need to plug the vehicle in to recharge the battery	Not Important	70	13.4%	43	8.5%	78	7.6%	11	6.1%
	Important	257	49.0%	322	63.5%	646	63.2%	122	67.8%
	Very Important	197	37.6%	142	28.0%	298	29.2%	47	26.1%
Concerns about limited access to plug-in locations	Not Important	42	8.1%	35	6.9%	170	16.7%	73	40.6%
	Important	236	45.6%	272	53.6%	500	49.0%	66	36.7%
	Very Important	239	46.2%	200	39.4%	350	34.3%	41	22.8%
Availability of desirable vehicle size or style	Not Important	98	18.7%	63	12.5%	84	8.2%	10	5.6%
	Important	243	46.4%	208	41.1%	537	52.7%	115	63.9%
	Very Important	183	34.9%	235	46.4%	398	39.1%	55	30.6%
Reliability of the vehicle	Not Important	34	6.5%	50	9.9%	159	15.6%	61	33.9%
	Important	221	42.4%	228	45.0%	489	48.0%	71	39.4%
	Very Important	266	51.1%	229	45.2%	370	36.3%	48	26.7%
On-going maintenance and operative costs	Not Important	37	7.1%	37	7.3%	50	4.9%	5	2.8%
	Important	236	45.4%	203	40.0%	514	50.4%	104	57.8%
	Very Important	247	47.5%	267	52.7%	456	44.7%	71	39.4%
The ability to carry heavy loads	Not Important	81	15.4%	47	9.3%	142	13.9%	58	32.2%
	Important	227	43.2%	228	45.0%	436	42.8%	71	39.4%
	Very Important	218	41.4%	232	45.8%	441	43.3%	51	28.3%
Cells highlighted represent the highest percentage across zones									

3.1.10 Impact on environment

Table 12 explains that more than 50% of the respondents from West India, North India, and East India agree that "cars, minivans, vans, pickups and SUVs are not an important source of air pollution anymore" while South India respondents disagree with this statement. 51.8% respondents from West India agree that "government rules allow minivans, vans, pickups and SUVs to pollute more than passenger cars, for every gallon of gas used" while 55.6% disagree with the statement from East India.

65.8% respondents from South India and 72.8% respondents from East India agreed that "cars, minivans, vans, pickups and SUVs are an important source of the greenhouse gases that many scientists believe are warming the earth's climate" while majority of the respondents from West India and North India neither agreed nor disagreed with that statement.

Respondents from all regions neither agreed nor disagreed with this statement: "Government rules require minivans, vans, pickups and SUVs to meet the same miles-per-gallon standards as passenger cars" in the majority. This clearly shows that they were not worried about government policies.

66.1% respondents from South India and 72.2% respondents from East India agreed that "exhaust from cars, minivans, vans, pickups and SUVs are an important source of pollution that cause asthma and make asthma attacks worse" while a majority of the respondents from West India and North India neither agreed nor disagreed with that statement.

Table 12: Breakup of impact on environment of various transport modes by region

Statement	Perceptions	Zone							
		South India		West India		North India		East India	
		Count	%	Count	%	Count	%	Count	%
"Cars, minivans, vans, pickups and SUVs are not an important source of air pollution anymore"	Agree	193	37.3%	433	84.2%	624	61.2%	129	71.7%
	Disagree	293	56.6%	50	9.7%	321	31.5%	47	26.1%
"Government rules allow minivans, vans, pickups and SUVs to pollute more than passenger cars, for every gallon of gas used"	Agree	226	44.2%	266	51.8%	390	38.3%	55	30.6%
	Disagree	165	32.3%	214	41.6%	440	43.3%	100	55.6%
"Cars, minivans, vans, pickups and SUVs are an important source of the greenhouse gases that many scientists believe are warming the earth's climate"	Agree	338	65.8%	236	45.9%	433	42.5%	131	72.8%
	Disagree	105	20.4%	155	30.2%	354	34.7%	36	20.0%
"Government rules require minivans, vans, pickups and SUVs to meet the same miles-per-gallon standards as passenger cars"	Agree	214	41.7%	174	33.9%	323	31.8%	61	33.9%
	Disagree	162	31.6%	216	42.0%	425	41.8%	89	49.4%
"Exhaust from cars, minivans, vans, pickups and SUVs is an important source of the pollution that causes asthma and makes asthma attacks worse"	Agree	343	66.1%	153	29.8%	395	38.8%	130	72.2%
	Disagree	83	16.0%	193	37.5%	380	37.3%	32	17.8%
Cells highlighted represent the highest percentage across zones									

3.1.11 Premium for electric vehicles

All respondents from all the regions were willing to pay a premium, ranging from 10% to 20%. The majority of the respondents from the South region and North region were willing to pay 15% premium for an electric vehicle. Respondents from the East region were willing to pay 10% premium and West region respondents were willing to pay 20% premium (Table 13).

Table 13: Willing to pay a premium for electric vehicle

		Zone							
		South India		West India		North India		East India	
		Count	Column N %	Count	Column N %	Count	Column N %	Count	Column N %
If very/somewhat interested in electric vehicles would you be willing to pay a premium to purchase an electric vehicle?	10%	148	29.9%	104	20.5%	284	28.6%	101	59.1%
	15%	215	43.4%	170	33.5%	484	48.8%	62	36.3%
	20%	132	26.7%	234	46.1%	224	22.6%	8	4.7%
Cells highlighted represent the highest percentage across zones									

3.2 Preferences based on city

To understand the perceptions based on different cities, we asked multiple questions to the respondents. The questions were asked to know the average daily travel, different types of information sources which influence their purchasing decisions, their views on purchasing electric vehicles, what they think about the role of the State and Central Government, important factors for purchasing any vehicle, different types of transport issues, various reasons that can influence their decisions to purchase electric vehicles in the future, impact of transport sector on the environment, especially vehicular pollution and finally the perception on the premium for electric vehicles. The results were analyzed by city and are summarized below.

3.2.1 Estimated average daily drive

It was revealed in Table 14 that the highest number of respondents who travel upto 10 km in a day, were from Hyderabad. Respondents from Manesar, UP travel up to 20 km in a day. This was because most respondents were living in Gurgaon and prefer to commute on a daily basis. Bengaluru had the highest number of respondents who travel up to 30 km in a day. Respondents from Pune shared that they travel up to 60 km in a day. Respondents from Mumbai travel more than 60 km in a day to complete their daily routine work. It can be inferred from the data that respondents from all the cities can be targeted for the electric vehicle, because the travel distance within the cities can be easily covered by electric vehicles.

Table 14: Estimatedaverage daily travel

Name of the City	Drive per day				
	Up to10 km	11 to 20 km	21 to 30 km	31-60 km	60 km or more
Bengaluru	3.2%	27.2%	39.2%	24.0%	6.4%
Chennai	28.2%	33.3%	15.4%	17.9%	5.1%
Kochi-II	46.6%	29.3%	15.5%	8.6%	0.0%
Coimbatore	35.8%	32.1%	13.2%	15.1%	3.8%
Faridabad	5.0%	52.5%	8.8%	17.5%	16.3%
Gurgaon	4.7%	58.8%	17.6%	16.5%	2.4%
Guwahati	41.9%	37.4%	8.4%	11.6%	.6%
Hyderabad	51.1%	29.5%	13.6%	3.4%	2.3%
Jaipur	5.9%	25.3%	13.5%	45.3%	10.0%
Kochi	35.0%	30.0%	15.0%	15.0%	5.0%
Lucknow	16.3%	21.1%	21.1%	26.4%	15.0%
Manesar	9.6%	54.2%	20.9%	11.9%	3.4%
Mumbai	4.7%	22.8%	26.9%	26.7%	18.9%
Pune	9.7%	39.4%	16.1%	32.9%	1.9%
Rohtak	14.0%	21.1%	22.6%	28.7%	13.6%
Thiruvananthapuram	34.5%	32.1%	20.2%	7.1%	6.0%
Cells highlighted represent the highest percentage across all the cities					

3.2.2 Information sources

Table 15 highlights the city wise preferences for various information sources. It was analyzed that television was the major resource for the Jaipur respondents, newspapers were the major information source for respondents from Faridabad. Radio was the major source of information for Manesar respondents. Pune respondents shared that their major information source is magazines. The internet was the major information source for Gurgaon respondents. In South Region, respondents from Kochi were more dependent on word of mouth for accessing information.

Table 15: Information sources for new technology

Name of the City	Television	Newspaper	Radio	Magazines	Internet	Word of mouth	Other
Bengaluru	56.0%	56.8%	6.4%	40.8%	67.2%	.8%	1.6%
Chennai	40.9%	50.0%	2.3%	31.8%	55.8%	27.3%	15.9%
Kochi-II	57.9%	50.0%	2.7%	32.9%	53.9%	35.5%	9.2%
Coimbatore	59.1%	31.8%	0.0%	4.5%	42.4%	9.1%	12.1%
Faridabad	58.8%	86.3%	22.5%	37.5%	77.5%	10.0%	5.1%
Gurgaon	57.6%	88.2%	22.4%	31.8%	92.9%	5.9%	3.5%
Guwahati	71.6%	64.5%	19.4%	34.2%	36.1%	23.9%	1.9%
Hyderabad	47.1%	42.6%	3.7%	16.2%	65.4%	14.7%	8.1%
Jaipur	71.8%	76.3%	31.2%	47.6%	82.9%	2.4%	22.4%
Kochi	66.7%	25.0%	0.0%	25.0%	62.5%	16.7%	12.5%
Lucknow	43.7%	34.4%	8.9%	30.0%	36.4%	8.9%	7.3%
Manesar	49.2%	84.2%	35.6%	32.2%	81.9%	10.7%	3.4%
Mumbai	58.1%	54.4%	26.9%	43.6%	48.3%	7.2%	13.1%
Pune	64.5%	52.3%	3.9%	50.3%	83.9%	1.9%	10.3%
Rohtak	64.9%	69.4%	18.9%	31.7%	68.3%	4.9%	4.9%
Thiruvananthapuram	65.6%	57.3%	7.3%	40.6%	68.8%	35.4%	11.5%
Cells highlighted represent the highest percentage across all the cities							

3.2.3 Influence on purchasing decisions

Table 16 highlights that in each district, the level of influence for purchasing a new vehicle is different. In Guwahati, elders take the decision for most of the respondents, in Faridabad decision is influenced by children and in Kochi, joint decision is taken for vehicle purchase.

Table 16: Vehicle purchase decision – influence sources

Name of the City	Who influences your vehicle purchase decision		
	Elders	Children	Joint decision
Bengaluru	51.2%	11.2%	37.6%
Chennai	20.5%	4.5%	75.0%
Kochi-II	42.3%	1.4%	56.3%
Coimbatore	50.0%	6.5%	43.5%
Faridabad	56.3%	31.3%	12.5%
Gurgaon	51.8%	22.4%	25.9%
Guwahati	81.3%	8.4%	10.3%
Hyderabad	56.6%	.9%	42.5%
Jaipur	49.4%	19.4%	31.2%
Kochi	18.2%	0.0%	81.8%
Lucknow	44.2%	9.6%	46.3%
Manesar	41.2%	26.6%	32.2%
Mumbai	58.1%	13.9%	28.1%
Pune	49.0%	14.8%	36.1%
Rohtak	61.9%	10.9%	27.2%
Thiruvananthapuram	35.1%	3.2%	61.7%
Cells highlighted represent the highest percentage across all the cities			

3.2.4 About new vehicles

Table 17 reveals that respondents from Guwahati are ready to purchase the vehicle once it is available in the market. Respondents from Pune will wait to read reviews of it and then buy it if the reviews are favorable. Respondents from Manesar and Faridabad will prefer to wait until the new technology is widely accepted and proven in the market before considering it.

Table 17: When a new vehicle becomes available for purchase, what do you do?

Name of the City	When a new vehicle becomes available for purchase, what do you do?			
	I am among the first to purchase it.	**I wait to read a review of it and then buy it if the review is favorable**	**I wait until this new technology has been widely accepted and proven before considering it.**	**Other.**
Bengaluru	25.2%	64.2%	9.8%	.8%
Chennai	4.4%	28.9%	62.2%	4.4%
Kochi-II	0.0%	44.2%	54.5%	1.3%
Coimbatore	13.4%	29.9%	46.3%	10.4%
Faridabad	7.5%	22.5%	66.3%	3.8%
Gurgaon	12.9%	37.6%	48.2%	1.2%
Guwahati	49.0%	27.1%	23.9%	0.0%
Hyderabad	1.5%	46.6%	45.9%	6.0%
Jaipur	8.8%	34.1%	51.8%	5.3%
Kochi	0.0%	59.1%	31.8%	9.1%
Lucknow	7.9%	51.5%	37.0%	3.5%
Manesar	6.2%	24.3%	66.1%	3.4%
Mumbai	9.2%	54.2%	35.3%	1.4%
Pune	14.0%	68.0%	16.0%	2.0%
Rohtak	30.9%	46.0%	18.9%	4.2%
Thiruvananthapuram	4.2%	41.7%	52.1%	2.1%
Cells highlighted represent the highest percentage across all the cities				

3.2.5 Interested in new electric vehicles

Respondents from all the cities shared that they are ready to purchase an electric vehicle once they become easily available in the market in the next couple of years. It was revealed in Table 18 that respondents from Guwahati and Gurgaon were "interested" and "somewhat interested" in purchasing an electric vehicle in the near future. Respondents from small cities were not very much interested in purchasing electric vehicles (Table 18).

Table 18: Interest level in purchasing electric vehicles

Name of the City	How interested would you be in purchasing an electric motor vehicle once they become easily available in the next couple of years?				
	Very interested	Somewhat interested	Not very interested	Not planning to purchase future vehicles	Cannot say
Bengaluru	24.0%	40.8%	12.8%	12.8%	9.6%
Chennai	51.1%	33.3%	6.7%	2.2%	6.7%
Kochi-II	25.0%	47.4%	11.8%	5.3%	10.5%
Coimbatore	32.8%	46.9%	9.4%	3.1%	7.8%
Faridabad	26.3%	56.3%	8.8%	5.0%	3.8%
Gurgaon	14.1%	61.2%	12.9%	9.4%	2.4%
Guwahati	65.8%	25.8%	2.6%	5.2%	.6%
Hyderabad	46.6%	42.9%	4.5%	2.3%	3.8%
Jaipur	18.2%	55.9%	7.6%	13.5%	4.7%
Kochi	28.0%	36.0%	20.0%	8.0%	8.0%
Lucknow	28.4%	35.0%	12.3%	9.9%	14.4%
Manesar	19.2%	50.3%	11.9%	16.4%	2.3%
Mumbai	24.4%	37.6%	15.7%	11.5%	10.7%
Pune	25.8%	27.2%	22.5%	12.6%	11.9%
Rohtak	22.6%	34.3%	23.4%	8.3%	11.3%
Thiruvananthapuram	34.4%	45.8%	4.2%	1.0%	14.6%
Cells highlighted represent the highest percentage across all the cities					

3.2.6 Role of State and Central Government

Respondents from all the cities shared that the transport policy of the State and Central Government should encourage electric vehicles. Respondents from Chennai and Lucknow were strongly in favor of encouraging electric vehicles for both public and private transport. Respondents from Guwahati were in favor of promoting the EVs only for public transport. Respondents from Faridabad shared that the transport policy of the State and Central Government for encouraging electric vehicles is not sufficient as of now.

Table 19: Encouragement for electric vehicles by State and Central Government

Name of the City	Do you think transport policy of the State and Central Government should encourage electric vehicles?		
	Yes, but only for public transport	**Yes, both for public and private transport**	**No, not worth it**
Bengaluru	32.8%	52.8%	14.4%
Chennai	22.7%	77.3%	0.0%
Kochi-II	18.9%	70.3%	10.8%
Coimbatore	30.2%	63.5%	6.3%
Faridabad	43.8%	20.0%	36.3%
Gurgaon	17.6%	67.1%	15.3%
Guwahati	74.0%	12.3%	13.6%
Hyderabad	18.9%	75.0%	6.1%
Jaipur	18.8%	69.4%	11.8%
Kochi	28.0%	64.0%	8.0%
Lucknow	12.4%	76.8%	10.8%
Manesar	48.0%	27.1%	24.9%
Mumbai	22.0%	62.8%	15.2%
Pune	33.8%	52.3%	13.9%
Rohtak	48.7%	42.1%	9.2%
Thiruvananthapuram	12.6%	83.2%	4.2%
Cells highlighted represent the highest percentage across all the cities			

3.2.7 Transport issues

Table 20 highlights that respondents shared that traffic congestion that they experience while driving is a major problem for Thiruvananthapuram, for Mumbai respondents it is a problem and for Guwahati respondents it is not a problem. Respondents from Chennai shared that traffic noise that they hear at home, work or school is a major problem for them, for Guwahati respondents it was a problem, and for Bengaluru respondents it was not a problem. Respondents from Hyderabad shared that vehicle emissions that affect local air quality area major problem for them, for Jaipur respondents it was a problem, for Guwahati it was not a problem.

Table 21 highlights that for Bengaluru respondents, vehicle emissions that contribute to climate change arenot a problem, for Guwahati respondents it is a problem, for Hyderabad respondents it is a major problem. Respondents shared that they feel unsafe because of speeding traffic; for Kochi respondents it was a major problem, for Jaipur respondents it was a problem, however Gurgaon respondents shared that it is not a problem. Respondents from Pune shared that fuel import from foreign countries is not a problem for them, for Guwahati respondents it is a problem and for Bengaluru respondents it is a major problem.

Table 20: Transport issues based on cities

Name of the City	Traffic congestion that you experience while driving			Traffic noise that you hear at home, work or school			Vehicle emissions that affect local air quality		
	Not a Problem	Problem	Major Problem	Not a Problem	Problem	Major Problem	Not a Problem	Problem	Major Problem
Bengaluru	28.0%	48.8%	23.2%	20.0%	58.4%	21.6%	14.4%	53.6%	32.0%
Chennai	6.8%	29.5%	63.6%	2.3%	50.0%	47.7%	0.0%	31.8%	68.2%
Kochi-II	3.9%	47.4%	48.7%	5.2%	72.7%	22.1%	0.0%	35.5%	64.5%
Coimbatore	10.3%	58.8%	30.9%	4.3%	55.1%	40.6%	4.5%	38.8%	56.7%
Faridabad	43.8%	53.8%	2.5%	10.0%	60.0%	30.0%	31.3%	53.8%	15.0%
Gurgaon	67.1%	32.9%	0.0%	9.4%	77.6%	12.9%	52.9%	34.1%	12.9%
Guwahati	80.0%	20.0%	0.0%	2.6%	86.5%	11.0%	76.8%	18.1%	5.2%
Hyderabad	3.8%	37.4%	58.8%	3.7%	52.6%	43.7%	2.2%	28.1%	69.6%
Jaipur	13.5%	82.4%	4.1%	10.0%	56.5%	33.5%	8.2%	71.2%	20.6%
Kochi	0.0%	36.0%	64.0%	4.0%	52.0%	44.0%	0.0%	32.0%	68.0%
Lucknow	20.1%	51.4%	28.5%	7.6%	61.0%	31.3%	5.3%	45.9%	48.8%
Manesar	50.8%	45.2%	4.0%	11.9%	69.5%	18.6%	45.8%	39.5%	14.7%
Mumbai	36.7%	61.1%	2.2%	11.1%	70.8%	18.1%	12.2%	53.9%	33.9%
Pune	9.9%	53.3%	36.8%	9.2%	53.9%	36.8%	11.2%	43.4%	45.4%
Rohtak	34.7%	55.5%	9.8%	14.3%	58.9%	26.8%	13.6%	55.5%	30.9%
Thiruvananthapuram	2.1%	33.3%	64.6%	15.6%	56.3%	28.1%	0.0%	42.1%	57.9%
Cells highlighted represent the highest percentage across all the cities									

Table 21: Transport issues based on cities

Name of the City	Vehicle emissions that contribute to climate change			Unsafe communities because of speeding traffic			Importing much of our oil from foreign countries		
	Not a Problem	Problem	Major Problem	Not a Problem	Problem	Major Problem	Not a Problem	Problem	Major Problem
Bengaluru	16.8%	51.2%	32.0%	8.8%	42.4%	48.8%	7.2%	23.2%	69.6%
Chennai	4.5%	29.5%	65.9%	11.4%	34.1%	54.5%	11.4%	38.6%	50.0%
Kochi-II	2.6%	40.3%	57.1%	1.4%	45.9%	52.7%	10.7%	49.3%	40.0%
Coimbatore	4.5%	46.3%	49.3%	7.6%	54.5%	37.9%	12.3%	47.7%	40.0%
Faridabad	11.3%	58.8%	30.0%	26.3%	57.5%	16.3%	8.8%	60.0%	31.3%
Gurgaon	10.6%	67.1%	22.4%	52.9%	29.4%	17.6%	14.1%	63.5%	22.4%
Guwahati	3.2%	85.8%	11.0%	72.9%	19.4%	7.7%	5.2%	82.6%	12.3%
Hyderabad	1.5%	28.4%	70.1%	5.3%	61.1%	33.6%	11.3%	36.1%	52.6%
Jaipur	8.2%	45.3%	46.5%	4.1%	70.6%	25.3%	6.5%	39.4%	54.1%
Kochi	0.0%	40.0%	60.0%	0.0%	40.0%	60.0%	4.0%	56.0%	40.0%
Lucknow	8.6%	50.2%	41.2%	8.9%	53.8%	37.2%	13.8%	47.2%	39.0%
Manesar	10.7%	65.5%	23.7%	41.8%	39.0%	19.2%	10.2%	65.0%	24.9%
Mumbai	15.8%	50.3%	33.9%	11.9%	54.7%	33.3%	11.9%	41.7%	46.4%
Pune	10.5%	40.8%	48.7%	11.2%	50.0%	38.8%	21.7%	36.8%	41.4%
Rohtak	15.1%	49.1%	35.8%	12.1%	48.3%	39.6%	3.8%	42.6%	53.6%
Thiruvananthapuram	0.0%	38.5%	61.5%	2.1%	45.7%	52.1%	14.7%	48.4%	36.8%
Cells highlighted represent the highest percentage across all the cities									

3.2.8 Impact on environment

Table 22 highlights that saving money on the cost of operation is not important for the respondents of Guwahati, for Jaipur respondents it is an important factor and for Hyderabad respondents it is a very important factor. Reduced impact on the environment is a very important factor for the respondents from Kochi, it is an important factor for the respondents from Gurgaon, however, it is not an important factor for Bengaluru respondents.

Reduced dependence on gasoline is a very important factor for Mumbai and Thiruvananthapuram respondents, for Jaipur respondents it is an important factor, however, for Guwahati respondents it is not an important factor. Driving a vehicle with more advanced or innovative technology is very important for Bengaluru respondents, for Gurgaon respondents it is an important factor and for Kochi it is not an important factor.

As explained in Table 23, the common issues were "recharge stations, higher price, availability of desirable vehicle size, reliability, on-going maintenance, operating costs and the ability to carry occasionally heavy loads" which are restricting the demand for EV in the Indian market.

Table 22: City wise preferences for various attributes

Name of the City	Saving money on the cost of operation (using electricity rather than gasoline)			Reduced impact on the environment			Reduced dependence on gasoline			Driving a vehicle with more advanced or innovative technology		
	Not Important	Important	Very Important	Not Important	Important	Very Important	Not Important	Important	Very Important	Not Important	Important	Very Important
Bengaluru	32.0%	44.8%	23.2%	19.2%	45.6%	35.2%	10.4%	60.0%	29.6%	10.4%	40.8%	48.8%
Chennai	4.7%	53.5%	41.9%	4.9%	39.0%	56.1%	12.2%	48.8%	39.0%	9.1%	45.5%	45.5%
Kochi-II	1.4%	58.1%	40.5%	1.4%	43.2%	55.4%	4.1%	56.8%	39.2%	14.1%	52.1%	33.8%
Coimbatore	7.6%	53.0%	39.4%	6.2%	50.8%	43.1%	7.6%	59.1%	33.3%	9.2%	55.4%	35.4%
Faridabad	80.0%	17.5%	2.5%	10.0%	82.5%	7.5%	65.0%	26.3%	8.8%	10.0%	77.5%	12.5%
Gurgaon	88.2%	11.8%	0.0%	3.5%	94.1%	2.4%	68.2%	20.0%	11.8%	4.7%	83.5%	11.8%
Guwahati	93.5%	6.5%	0.0%	7.1%	92.9%	0.0%	86.5%	9.7%	3.9%	6.5%	91.6%	1.9%
Hyderabad	2.2%	53.7%	44.0%	1.5%	29.9%	68.7%	5.3%	57.1%	37.6%	9.0%	56.0%	35.1%
Jaipur	34.1%	62.9%	2.9%	11.8%	69.8%	18.3%	17.6%	62.4%	20.0%	6.5%	56.5%	37.1%
Kochi	16.0%	48.0%	36.0%	0.0%	28.0%	72.0%	0.0%	56.0%	44.0%	24.0%	32.0%	44.0%
Lucknow	23.4%	52.5%	24.2%	5.7%	58.7%	35.6%	10.5%	46.2%	43.3%	10.1%	50.2%	39.7%
Manesar	91.0%	9.0%	0.0%	7.9%	91.0%	1.1%	78.0%	15.8%	6.2%	7.3%	89.3%	3.4%
Mumbai	45.1%	53.5%	1.4%	6.8%	76.6%	16.6%	8.2%	46.2%	45.6%	11.3%	44.5%	44.2%
Pune	9.9%	56.6%	33.6%	5.3%	48.0%	46.7%	8.6%	52.0%	39.5%	12.6%	42.4%	45.0%
Rohtak	39.6%	49.8%	10.6%	18.5%	60.0%	21.5%	11.0%	54.9%	34.1%	8.7%	45.8%	45.5%
Thiruvananthapuram	2.1%	54.7%	43.2%	1.0%	40.6%	58.3%	3.1%	51.0%	45.8%	13.5%	50.0%	36.5%

Cells highlighted represent the highest percentage across all the cities

Table 23: Concerns about purchase of electric vehicles

Name of the City	A higher purchase price than for a comparable conventional vehicle			The need to plug the vehicle in to recharge the battery			Concerns about limited access to plug-in locations		
	Not Important	Important	Very Important	Not Important	Important	Very Important	Not Important	Important	Very Important
Bengaluru	16.0%	59.2%	24.8%	6.4%	56.8%	36.8%	7.2%	45.6%	47.2%
Chennai	16.3%	48.8%	34.9%	20.9%	27.9%	51.2%	4.8%	23.8%	71.4%
Kochi-II	11.6%	69.6%	18.8%	17.6%	47.1%	35.3%	4.5%	48.5%	47.0%
Coimbatore	22.6%	58.1%	19.4%	14.5%	48.4%	37.1%	12.9%	58.1%	29.0%
Faridabad	38.8%	52.5%	8.8%	8.8%	71.3%	20.0%	26.3%	52.5%	21.3%
Gurgaon	67.1%	27.1%	5.9%	9.4%	72.9%	17.6%	43.5%	37.6%	18.8%
Guwahati	56.8%	37.4%	5.8%	5.2%	70.3%	24.5%	46.5%	36.1%	17.4%
Hyderabad	14.1%	62.5%	23.4%	17.7%	46.9%	35.4%	10.3%	55.6%	34.1%
Jaipur	18.8%	64.7%	16.5%	7.6%	55.3%	37.1%	8.8%	57.1%	34.1%
Kochi	8.0%	52.0%	40.0%	12.0%	52.0%	36.0%	4.0%	40.0%	56.0%
Lucknow	24.7%	58.8%	16.5%	10.2%	57.6%	32.2%	5.8%	48.6%	45.7%
Manesar	45.8%	42.4%	11.9%	3.4%	74.6%	22.0%	33.3%	37.9%	28.8%
Mumbai	27.9%	61.7%	10.4%	8.2%	73.2%	18.6%	5.6%	56.6%	37.7%
Pune	7.2%	57.2%	35.5%	9.2%	40.8%	50.0%	9.9%	46.7%	43.4%
Rohtak	13.6%	70.6%	15.8%	7.2%	60.4%	32.5%	9.1%	54.3%	36.6%
Thiruvananthapuram	8.3%	51.0%	40.6%	9.4%	53.1%	37.5%	7.3%	32.3%	60.4%

Cells highlighted represent the highest percentage across all the cities

Table 24: Factors supporting purchase of electric vehicles

Name of the City	Availability of desirable vehicle size or style			Reliability of the vehicle			On-going maintenance and operative costs (including battery replacement)			The ability to carry heavy loads		
	Not Important	Important	Very Important	Not Important	Important	Very Important	Not Important	Important	Very Important	Not Important	Important	Very Important
Bengaluru	17.6%	31.2%	51.2%	11.2%	44.0%	44.8%	13.6%	40.0%	46.4%	16.8%	42.4%	40.8%
Chennai	21.4%	42.9%	35.7%	2.4%	40.5%	57.1%	4.8%	38.1%	57.1%	23.3%	48.8%	27.9%
Kochi-II	16.4%	53.7%	29.9%	8.8%	39.7%	51.5%	6.0%	49.3%	44.8%	11.8%	45.6%	42.6%
Coimbatore	11.1%	60.3%	28.6%	8.1%	45.2%	46.8%	12.9%	40.3%	46.8%	9.5%	47.6%	42.9%
Faridabad	5.0%	65.0%	30.0%	23.8%	50.0%	26.3%	5.0%	65.0%	30.0%	20.0%	52.5%	27.5%
Gurgaon	3.5%	64.7%	31.8%	37.6%	42.4%	20.0%	4.7%	63.5%	31.8%	34.1%	44.7%	21.2%
Guwahati	3.9%	65.2%	31.0%	39.4%	38.1%	22.6%	3.2%	61.9%	34.8%	34.2%	37.4%	28.4%
Hyderabad	19.1%	52.7%	28.2%	3.1%	43.8%	53.1%	2.3%	51.6%	46.1%	16.0%	42.0%	42.0%
Jaipur	5.9%	45.9%	48.2%	8.8%	54.1%	37.1%	5.9%	42.4%	51.8%	6.5%	48.8%	44.7%
Kochi	16.0%	56.0%	28.0%	0.0%	48.0%	52.0%	0.0%	32.0%	68.0%	20.0%	52.0%	28.0%
Lucknow	11.6%	55.8%	32.6%	10.0%	53.1%	36.9%	5.8%	49.8%	44.4%	10.7%	43.8%	45.5%
Manesar	3.4%	65.5%	31.1%	24.3%	42.4%	33.3%	3.4%	62.7%	33.9%	23.2%	43.5%	33.3%
Mumbai	12.7%	47.5%	39.8%	10.1%	48.5%	41.4%	7.6%	39.7%	52.7%	9.0%	42.5%	48.5%
Pune	11.8%	26.3%	61.8%	9.2%	36.8%	53.9%	6.6%	40.8%	52.6%	9.9%	50.7%	39.5%
Rohtak	12.5%	38.1%	49.4%	9.8%	44.5%	45.7%	4.5%	39.2%	56.2%	7.2%	34.0%	58.9%
Thiruvananthapuram	25.0%	44.8%	30.2%	4.2%	39.6%	56.3%	3.1%	47.9%	49.0%	15.6%	38.5%	45.8%
Cells highlighted represent the highest percentage across all the cities												

Table 25 explains that respondents from most cities agreed with the statement that cars, minivans, vans, pickups and SUVs are not an important source of air pollution anymore. Respondents from Kochi disagreed with the statement and shared that cars, minivans, vans, pickups and SUVs are an important source of air pollution.

Respondents from Coimbatore agreed with the statement that government rules allow minivans, vans, pickups and SUVs to pollute more than passenger cars, for every gallon of gas used, however respondents from Faridabad disagreed with that statement.

Respondents from Hyderabad agreed with the statement that cars, minivans, vans, pickups and SUVs are an important source of the greenhouse gases that many scientists believe are warming the earth's climate, however respondents from the Faridabad disagreed with this statement.

Respondents from Chennai agreed with the statement that exhaust from cars, minivans, vans, pickups and SUVs is an important source of the pollution that causes asthma and makes asthma attacks worse, however respondents from Faridabad disagreed with the above-mentioned statement. Bengaluru respondents were unsure about the above-mentioned statement.

Table 25: City wise breakup of impact on environment of various transport modes

Name of the City	Cars, minivans, vans, pickups and SUVs are not an important source of air pollution anymore			Government rules allow minivans, vans, pickups and SUVs to pollute more than passenger cars, for every gallon of gas used			Cars, minivans, vans, pickups and SUVs are an important source of the greenhouse gases that many scientists believe are warming the earth's climate			Exhaust from cars, minivans, vans, pickups and SUVs is an important source of the pollution that causes asthma and makes asthma attacks worse		
	Agree	Disagree	Unsure	Agree	Disagree	Unsure	Agree	Disagree	Unsure	Agree	Disagree	Unsure
Bengaluru	84%	10%	6%	37%	54%	10%	33%	44%	23%	28%	31%	41%
Chennai	28%	70%	2%	39%	27%	34%	74%	16%	9%	84%	12%	5%
Kochi-II	17%	78%	5%	44%	30%	25%	73%	14%	13%	74%	9%	17%
Coimbatore	49%	43%	8%	68%	22%	10%	71%	17%	12%	59%	22%	19%
Faridabad	40%	59%	1%	10%	63%	28%	34%	49%	18%	30%	53%	18%
Gurgaon	68%	28%	4%	44%	44%	13%	44%	41%	15%	39%	40%	21%
Guwahati	82%	17%	1%	28%	60%	12%	75%	19%	5%	74%	18%	8%
Hyderabad	15%	77%	8%	49%	24%	27%	80%	11%	9%	83%	9%	8%
Jaipur	70%	19%	11%	53%	26%	21%	45%	28%	26%	39%	29%	32%
Kochi	8%	84%	8%	44%	28%	28%	56%	24%	20%	60%	16%	24%
Lucknow	50%	42%	8%	46%	32%	21%	54%	16%	30%	61%	20%	18%
Manesar	49%	45%	6%	28%	51%	21%	42%	43%	15%	34%	50%	16%
Mumbai	87%	7%	7%	56%	38%	6%	45%	28%	28%	26%	39%	35%
Pune	79%	17%	5%	41%	51%	8%	49%	36%	15%	38%	35%	27%
Rohtak	79%	13%	8%	36%	53%	11%	33%	45%	22%	23%	45%	32%
Thiruvananthapuram	17%	77%	6%	35%	25%	40%	77%	10%	13%	83%	8%	8%

Cells highlighted represent the highest percentage across all the cities

3.2.9 Premium for electric vehicles

Table 26 highlights that respondents from all cities were willing to pay a premium for electric vehicles. It was revealed that respondents from Guwahati were ready to pay 10% premium over the price of a conventional vehicle to save the environment, Gurgaon respondents were ready to pay 15% premium and Mumbai respondents were ready to pay 20% premium.

Table 26: Willing to pay apremium for electric vehicle

Name of the City	If very/somewhat interested in electric vehicles would you be willing to pay a premium to purchase anelectric vehicle?		
	10% Premium	15% Premium	20% Premium
Bengaluru	16.3%	42.3%	41.5%
Chennai	39.0%	36.6%	24.4%
Kochi-II	25.0%	50.0%	25.0%
Coimbatore	30.4%	33.9%	35.7%
Faridabad	10.0%	70.0%	20.0%
Gurgaon	25.9%	72.9%	1.2%
Guwahati	59.6%	38.4%	2.1%
Hyderabad	33.6%	47.9%	18.5%
Jaipur	43.2%	41.4%	15.4%
Kochi	56.0%	24.0%	20.0%
Lucknow	31.1%	37.8%	31.1%
Manesar	12.1%	68.4%	19.5%
Mumbai	18.0%	31.5%	50.6%
Pune	26.3%	38.2%	35.5%
Rohtak	34.7%	35.5%	29.7%
Thiruvananthapuram	42.4%	43.5%	14.1%
Cells highlighted represent the highest percentage across all the cities			

3.3 Preferences based on age groups

To understand the perceptions based on age groups, we asked multiple questions to the respondents. The questions were asked to know the average daily travel, different types of information sources, who influences their purchasing decisions, their views on purchasing electric vehicles, what they think about the role of the State and Central Government, important factors for purchasing any vehicle, different types of transport issues, various reasons which can influence their decisions to purchase an electric vehicle in the future, impact of transport sector on the environment, especially vehicular pollution and finally the perception on the premium for electric vehicles. The results were analyzed age group wise and are summarized below.

3.3.1 Estimated average daily drive

Table 27 highlights that the age group of 18-30 and 31-60 travel up to 20 km per day for their routine work, however, the age group of above 61 years travels up to 60 km or more in a day. Considering the data, it is clear that the age group of 18-30 years, if motivated properly and given sufficient services can think of using electric vehicles for their short travels.

Table 27: Drive per day based on age group

		Age group of the respondents					
		18-30		31-60		61 and Above	
		Count	Column N %	Count	Column N %	Count	Column N %
Drive per day	Up to10 km	266	23.2%	61	7.4%	14	9.9%
	11 to 20 km	367	32.1%	268	32.4%	31	22.0%
	21 to 30 km	221	19.3%	184	22.2%	19	13.5%
	31-60 km	206	18.0%	243	29.4%	37	26.2%
	60 km or more	85	7.4%	71	8.6%	40	28.4%

Cells highlighted represent the highest percentage across age groups

3.3.2 Source of information

Table 28reveals that television is an important source of information among the age groups of 18-30 and above 61 years, however, among the age group of 31-60, newspapers are the main source of information followed by the internet.

Table 28: Source of information for new technology

Source of information	Age group of the respondents					
	18-30		31-60		61 and above	
	Count	Column N %	Count	Column N %	Count	Column N %
Television	768	61.2%	488	58.9%	84	59.6%
Newspaper	695	55.4%	562	67.9%	65	46.1%
Radio	151	12.1%	206	24.8%	21	14.9%
Magazines	418	33.3%	318	38.4%	55	39.0%
Internet	739	59.0%	554	67.0%	44	31.2%
Word of mouth	154	12.3%	64	7.7%	14	9.9%
Cells highlighted represent the highest percentage across age groups						

3.3.3 Influence on purchasing decision

Table 29 reveals that age groups 18-30 and 31-60 discuss the decision with their elders while taking a decision about purchasing any vehicle. However respondents from the age group of 61 and above take joint decisions while purchasing any vehicle, this group also discusses it with children while making any purchase decisions.

Table 29: Influences your vehicle purchase decision

	Age group of the respondents					
	18-30		31-60		61 and above	
	Count	Column N %	Count	Column N %	Count	Column N %
Elders	647	53.4%	455	54.9%	45	31.9%
Children	112	9.2%	132	15.9%	43	30.5%
Joint decision	453	37.4%	242	29.2%	53	37.6%
Cells highlighted represent the highest percentage across age groups						

3.3.4 About new vehicles

In Table 30, respondents from all age groups shared that they prefer to wait to read the reviews of a new vehicle and then buy it if the reviews are favorable.

Table 30: When a new vehicle becomes available for purchase, what do you do?

Decision	Age group of the respondents					
	18-30		31-60		61 and Above	
	Count	%	Count	%	Count	%
I am among the first to purchase it.	195	15.8%	111	13.5%	13	9.3%
I wait to read a review of it and then buy it if the review is favorable.	567	45.9%	349	42.5%	62	44.3%
I wait until this new technology has been widely accepted and proven before considering it.	440	35.7%	334	40.7%	59	42.1%
Other.	32	2.6%	27	3.3%	6	4.3%
Cells highlighted represent the highest percentage across age groups						

3.3.5 Interested in new electric vehicle

Table 31 highlights that respondents from all age groups are interested in buyingan electric vehicle once they become easily available in the next couple of years. The age group 18-30 has higher percentage of respondents compared to the other age groups who are very interested in buying such vehicles.

Table 31: Interested in purchasing electric vehicle

		Age group of the respondents					
		18-30		31-60		61 and above	
		Count	Column N %	Count	Column N %	Count	Column N %
How interested would you be in purchasing an electric motor vehicle once they become easily available in the next couple of years?	Very interested	421	33.8%	194	23.6%	24	17.3%
	Somewhat interested	480	38.5%	371	45.1%	45	32.4%
	Not very interested	136	10.9%	118	14.3%	28	20.1%
	Not planning to purchase future vehicles	99	7.9%	83	10.1%	23	16.5%
	Cannot say	110	8.8%	57	6.9%	19	13.7%
Cells highlighted represent the highest percentage across age groups							

3.3.6 Role of State and Central Government

Table 32 revealed that respondents from all age groups felt that the State and Central Government should encourage electric vehicles for both public and private transport by making amendments to the transport policy. Approximately 25% of the respondents who were above 61 years felt that the policies defined by the State and Central Government are not sufficient for both the sectors.

Table 32: Encouragement of electric vehicles by State and Central Government

		Age group of the respondents					
		18-30		31-60		61 and above	
		Count	Column N %	Count	Column N %	Count	Column N %
Do you think transport policy of the State and Central Government should encourage electric vehicles?	Yes, but only for public transport	363	29.3%	288	35.2%	35	25.2%
	Yes, both for public and private transport	748	60.5%	400	48.8%	69	49.6%
	No, not worth it	126	10.2%	131	16.0%	35	25.2%
Cells highlighted represent the highest percentage across age groups							

3.3.7 Transport issues

Table 33 shows that factors such as fuel efficiency, vehicle power, reliability and fuel type are very important for all the age groups. Safety is a very important factor for the age group of 18-30, but important to the other two age groups. Purchase price, reputation of the vehicle, vehicle emissions and pollution are important factors for all the age groups.

Table 33: Transport issues based on age groups

		Age group of the respondents					
		18-30		31-60		61 and above	
		Count	Column N %	Count	Column N %	Count	Column N %
Fuel efficiency	Very Important	1079	85.6%	666	80.3%	97	69.3%
	Important	172	13.6%	149	18.0%	41	29.3%
	Not Important	10	.8%	14	1.7%	2	1.4%
Safety	Very Important	874	69.3%	401	48.4%	68	48.2%
	Important	379	30.1%	407	49.1%	72	51.1%
	Not Important	8	.6%	21	2.5%	1	.7%
Vehicle power	Very Important	817	64.9%	520	62.7%	81	57.4%
	Important	406	32.3%	249	30.0%	51	36.2%
	Not Important	35	2.8%	60	7.2%	9	6.4%
Purchase price	Very Important	576	45.7%	338	40.8%	53	37.9%
	Important	651	51.6%	456	55.0%	81	57.9%
	Not Important	34	2.7%	35	4.2%	6	4.3%
Reliability	Very Important	775	62.0%	509	61.4%	83	58.9%
	Important	437	34.9%	276	33.3%	52	36.9%
	Not Important	39	3.1%	44	5.3%	6	4.3%
Vehicle size (to accommodate passengers or cargo)	Very Important	366	29.2%	262	31.6%	56	40.0%
	Important	792	63.3%	514	61.9%	73	52.1%
	Not Important	94	7.5%	54	6.5%	11	7.9%
Expected operating costs (for maintenance and repair)	Very Important	579	46.1%	404	48.7%	75	53.2%
	Important	614	48.9%	359	43.3%	58	41.1%
	Not Important	62	4.9%	66	8.0%	8	5.7%
Reputation of particular vehicle make or model	Very Important	389	31.4%	233	28.1%	45	31.9%
	Important	731	59.0%	543	65.5%	84	59.6%
	Not Important	118	9.5%	53	6.4%	12	8.5%
Fuel type (e.g. petrol, diesel, CNG)	Very Important	617	49.2%	395	47.6%	76	53.9%
	Important	519	41.4%	376	45.4%	55	39.0%
	Not Important	118	9.4%	58	7.0%	10	7.1%
Vehicle emissions and pollution	Very Important	546	43.5%	235	28.3%	44	31.2%
	Important	607	48.4%	541	65.2%	85	60.3%
	Not Important	102	8.1%	54	6.5%	12	8.5%
Cells highlighted represent the highest percentage across age groups							

3.3.8 Various reasons for purchasing electric vehicles in future

Questions relating to multiple factors were asked to understand the perceptions of different age groups. The factors were: traffic congestion that they experience while driving, traffic noise that they hear at home, work or school, vehicle emissions that affect local air quality, vehicle emissions that contribute to climate change, unsafe communities because of speeding traffic and importing much of our oil from foreign countries. We found that these factors are a problem for the majority of the respondents (Table 34).

Table 34: Age group wise perceptions

		Age group of the respondents					
		18-30		31-60		61 and above	
		Count	Column N %	Count	Column N %	Count	Column N %
Traffic congestion that you experience while driving.	Not a Problem	348	27.8%	266	32.1%	55	39.0%
	Problem	590	47.2%	461	55.6%	70	49.6%
	Major Problem	312	25.0%	102	12.3%	16	11.3%
Traffic noise that you hear at home, work or school	Not a Problem	128	10.2%	73	8.8%	21	14.9%
	Problem	796	63.4%	536	64.7%	85	60.3%
	Major Problem	332	26.4%	220	26.5%	35	24.8%
Vehicle emissions that affect local air quality	Not a Problem	222	17.8%	165	19.9%	29	20.6%
	Problem	545	43.6%	407	49.1%	71	50.4%
	Major Problem	482	38.6%	257	31.0%	41	29.1%
Vehicle emissions that contribute to climate change	Not a Problem	105	8.4%	95	11.5%	20	14.2%
	Problem	609	48.7%	447	53.9%	79	56.0%
	Major Problem	536	42.9%	287	34.6%	42	29.8%
Unsafe communities because of speeding traffic	Not a Problem	211	16.9%	165	19.9%	26	18.4%
	Problem	594	47.7%	407	49.1%	78	55.3%
	Major Problem	440	35.3%	257	31.0%	37	26.2%
Importing much of our oil from foreign countries	Not a Problem	143	11.5%	73	8.8%	18	12.8%
	Problem	580	46.5%	408	49.2%	66	46.8%
	Major Problem	523	42.0%	348	42.0%	57	40.4%
Cells highlighted represent the highest percentage across age groups							

Table 35 highlights that saving money on the cost of operation (using electricity rather than gasoline) is not an important factor for the age group of 31-60 and people above 61 years, however, it is an important factor for the age group of 18-30. Respondents from all the age groups were concerned about the environmental impact and highlighted that electric vehicles will reduce pollution in the coming future. For the age group of 18-30 and 31-60, it is important to reduce the dependence on gasoline, however, respondents above 61 years do not consider it an important factor. Respondents from all the age groups shared that "it is important to drive a vehicle with more advanced or innovative technology".

Table 35: Preferences for various attributes by age group

		Age group of the respondents					
		18-30		31-60		61 and above	
		Count	Column N %	Count	Column N %	Count	Column N %
Saving money on the cost of operation (using electricity rather than gasoline)	Not Important	403	32.4%	414	50.2%	79	56.8%
	Important	593	47.6%	327	39.6%	47	33.8%
	Very Important	249	20.0%	84	10.2%	13	9.4%
Reduced impact on the environment	Not Important	101	8.1%	65	7.9%	17	12.2%
	Important	716	57.5%	605	73.4%	103	74.1%
	Very Important	429	34.4%	154	18.7%	19	13.7%
Reduced dependence on gasoline	Not Important	269	21.6%	221	26.8%	54	38.8%
	Important	592	47.5%	342	41.5%	48	34.5%
	Very Important	385	30.9%	262	31.8%	37	26.6%
Driving a vehicle with more advanced or innovative technology	Not Important	123	9.9%	78	9.5%	13	9.4%
	Important	640	51.4%	523	63.5%	92	66.2%
	Very Important	481	38.7%	223	27.1%	34	24.5%
Cells highlighted represent the highest percentage across age groups							

As explained in Table 36, variables such as purchase price, recharge options, recharging stations, availability of vehicle size and vehicle reliability are important factors for respondents from all age groups. On-going maintenance and operative costs and load capacity are very important factors for the age group of 18-30 however these factors are important for rest of the age groups.

Table 36: Factors supporting purchase of electric vehicles by age group

		Age group of the respondents					
		18-30		31-60		61 and above	
		Count	Column N %	Count	Column N %	Count	Column N %
A higher purchase price than for a comparable conventional vehicle	Not Important	288	23.3%	225	27.3%	51	36.7%
	Important	731	59.2%	451	54.7%	64	46.0%
	Very Important	215	17.4%	148	18.0%	24	17.3%
The need to plug the vehicle in to recharge the battery	Not Important	126	10.2%	64	7.8%	9	6.5%
	Important	738	59.6%	500	60.7%	92	66.2%
	Very Important	374	30.2%	260	31.6%	38	27.3%
Concerns about limited access to plug-in locations	Not Important	177	14.4%	116	14.1%	25	18.0%
	Important	609	49.5%	390	47.3%	60	43.2%
	Very Important	444	36.1%	318	38.6%	54	38.8%
Availability of desirable vehicle size or style	Not Important	161	13.1%	74	9.0%	15	10.8%
	Important	600	48.7%	419	50.9%	70	50.4%
	Very Important	472	38.3%	330	40.1%	54	38.8%
Reliability of the vehicle	Not Important	159	12.9%	114	13.8%	28	20.1%
	Important	556	45.2%	381	46.2%	61	43.9%
	Very Important	516	41.9%	329	39.9%	50	36.0%
On-going maintenance and operative costs (including battery replacement)	Not Important	71	5.8%	46	5.6%	10	7.2%
	Important	571	46.3%	401	48.7%	70	50.4%
	Very Important	590	47.9%	377	45.8%	59	42.4%
The ability to carry heavy loads	Not Important	199	16.1%	94	11.4%	30	21.6%
	Important	510	41.3%	377	45.8%	61	43.9%
	Very Important	526	42.6%	353	42.8%	48	34.5%
Cells highlighted represent the highest percentage across age groups							

3.3.9 Impact on environment

Table 37 explains that respondents from all the age groups agreed with the statement that "cars, minivans, vans, pickups and SUVs are not an important source of air pollution anymore". Respondents from the age group of 18-30 and above 61 years agreed with the statement that government rules allow minivans, vans, pickups and SUVs to pollute more than passenger cars, for every gallon of gas used, however respondents from the age group of 31-60 disagree with the statement.

Respondents from all the age groups agreed with the statement that cars, minivans, vans, pickups and SUVs are an important source of greenhouse gases that many scientists believe are warming the earth's climate. Respondents from all the age groups disagreed with the statement that government rules require minivans, vans, pickups and SUVs to meet the same miles-per-gallon standards as passenger cars.

Respondents from the age groups 18-30 and above 61 years agreed with the statement that exhaust from cars, minivans, vans, pickups and SUVs is an important source of the pollution that causes asthma and makes asthma attacks worse, however the age group of 31-60 disagreed with the above-mentioned statement.

Table 37: Breakup of impact on environment of various transport modes by age group

		Age group of the respondents					
		18-30		31-60		61 and above	
		Count	Column N %	Count	Column N %	Count	Column N %
Cars, minivans, vans, pickups and SUVs are not an important source of air pollution anymore.	Agree	713	58.0%	561	67.7%	91	65.5%
	Disagree	448	36.5%	213	25.7%	33	23.7%
	Unsure	68	5.5%	55	6.6%	15	10.8%
Government rules allow minivans, vans, pickups and SUVs to pollute more than passenger cars, for every gallon of gas used.	Agree	554	45.3%	313	37.8%	58	42.0%
	Disagree	478	39.1%	379	45.7%	51	37.0%
	Unsure	191	15.6%	137	16.5%	29	21.0%
Cars, minivans, vans, pickups and SUVs are an important source of the greenhouse gases that many scientists believe are warming the earth's climate.	Agree	709	57.8%	343	41.4%	59	42.4%
	Disagree	332	27.1%	268	32.4%	45	32.4%
	Unsure	185	15.1%	217	26.2%	35	25.2%
Government rules require minivans, vans, pickups and SUVs to meet the same miles-per-gallon standards as passenger cars.	Agree	480	39.2%	229	27.6%	44	31.9%
	Disagree	485	39.7%	344	41.5%	55	39.9%
	Unsure	258	21.1%	256	30.9%	39	28.3%
Exhaust from cars, minivans, vans, pickups and SUVs is an important source of the pollution that causes asthma and makes asthma attacks worse.	Agree	676	55.0%	269	32.4%	57	41.0%
	Disagree	320	26.0%	315	38.0%	44	31.7%
	Unsure	233	19.0%	245	29.6%	38	27.3%
Cells highlighted represent the highest percentage across age groups							

3.3.10 Premium for electric vehicles

Respondents from all the age groups were willing to pay a premium for electric vehicles. It was revealed that respondents were ready to pay 15% premium over the conventional vehicle to save the environment (Table 38).

Table 38: Willing to give premium for electric vehicle

		Age group of the respondents					
		18-30		31-60		61 and above	
		Count	Column N %	Count	Column N %	Count	Column N %
If very/somewhat interested in electric vehicles would you be willing to pay a premium to purchase an electric vehicle?	10%	392	33.1%	209	25.6%	21	15.6%
	15%	447	37.7%	403	49.4%	71	52.6%
	20%	347	29.3%	203	24.9%	43	31.9%
Cells highlighted represent the highest percentage across age groups							

3.4 Preferences based on gender

To understand the perceptions based on gender, we asked multiple questions to the respondents. The questions were asked to know the average daily travel, different types of information sources, who influences their purchasing decisions, their views on purchasing an electric vehicle, what they think about the role of State and Central Governments, important factors for purchasing any vehicle, different types of transport issues, various reasons which can influence their decisions to purchase an electric vehicle in the future, impact of transport sector on the environment, especially vehicular pollution and finally the perception on the premium for electric vehicles. The results were analyzed gender wise and are summarized below.

3.4.1 Estimated average daily drive

Table 39 highlights that there is no significant difference between travel distance covered by male and female respondents. It is evident from the table that approximately 71% of male respondents and 63% of female respondents travel upto 30 km in a day for their routine work.

Table 39: Estimated average daily travel

		Gender			
		Male		Female	
		Count	Column N %	Count	Column N %
Drive per day	Up to10 km	227	18.8%	114	12.6%
	11 to 20 km	400	33.1%	271	29.9%
	21 to 30 km	236	19.5%	187	20.6%
	31-60 km	246	20.3%	236	26.0%
	60 km or more	100	8.3%	98	10.8%
Cells highlighted represent the highest percentage across genders					

3.4.2 Information sources

Table 40 highlights that there is no significant difference which exists in the information sources for both the genders. The main source of information for male and female respondents is the internet. Television followed by newspapers and magazines were the other important information sources for male respondents. On the other side, newspapers followed by television and magazines were the other important information sources for female respondents.

Table 40: Information sources for new technologies

	Gender			
	Male		Female	
	Count	Column N %	Count	Column N %
Television	736	58.5%	557	57.6%
Newspaper	712	56.6%	608	62.9%
Radio	187	14.9%	191	19.8%
Magazines	425	33.8%	364	37.6%
Internet	769	61.1%	620	64.2%
Word of mouth	156	12.4%	82	8.5%
Other	109	8.7%	86	8.9%
Cells highlighted represent the highest percentage across gender				

3.4.3 Influence on purchasing decision

It is clear from Table 41 that elders influence the purchasing decisions for the both the genders. However, women respondents give more priority to children and would like to go for a joint decision compared to male respondents.

Table 41: Vehicle purchase decision – influence source

		Gender			
		Male		Female	
		Count	Column N %	Count	Column N %
Who influences your vehicle purchase decision?	Elders	668	54.0%	482	51.0%
	Children	139	11.2%	144	15.2%
	Joint decision	431	34.8%	320	33.8%
Cells highlighted represent the highest percentage across genders					

3.4.4 About new vehicles

Table 42 highlights that both male and female respondents wanted to wait for the reviews of the vehicle and then buy it if the reviews are favorable.

Table 42: When a new vehicle becomes available for purchase, what do you do?

		Gender			
		Male		Female	
		Count	Column N %	Count	Column N %
When a new vehicle becomes available for purchase, what do you do?	I am among the first to purchase it.	204	16.5%	115	12.0%
	I wait to read a review of it and then buy it if the review is favorable.	545	44.0%	426	44.4%
	I wait until this new technology has been widely accepted and proven before considering it.	451	36.4%	389	40.6%
	Other.	40	3.2%	29	3.0%
Cells highlighted represent the highest percentage across genders					

3.4.5 Interested in new electric vehicle

Respondents were ready to purchase an electric vehicle once they become easily available in the next couple of years. Female respondents were more likely to buy such vehicles compared to male respondents (Table 43).

Table 43: Interested in purchasing electric vehicle

		Gender			
		Male		Female	
		Count	Column N %	Count	Column N %
How interested would you be in purchasing an electric motor vehicle once they become easily available in the next couple of years?	Very interested	391	31.3%	247	25.7%
	Somewhat interested	475	38.0%	429	44.6%
	Not very interested	161	12.9%	121	12.6%
	Not planning to purchase future vehicles	102	8.2%	101	10.5%
	Cannot say	120	9.6%	64	6.7%
Cells highlighted represent the highest percentage across genders					

3.4.6 Role of State and Central Government

Both male and female respondents shared that the State and Central Government should encourage transport policy for electric vehicles (Table 44).

Table 44: Encouragement of electric vehicles by State and Central Government

		Gender			
		Male		**Female**	
		Count	**Column N %**	**Count**	**Column N %**
Do you think transport policy of the State and Central Government should encourage electric vehicles?	Yes, but only for public transport	427	34.4%	258	27.0%
	Yes, both for public and private transport	642	51.7%	575	60.2%
	No, not worth it	173	13.9%	122	12.8%
Cells highlighted represent the highest percentage across genders					

3.4.7 Transport issues

Table 45 shows that female respondents were more concerned about traffic congestions than male respondents were. They expressed that traffic congestions are problematic for them. Traffic noise was a problem for both male and female respondents. Male respondents expressed that vehicle emissions which affect local air quality are a problem for them.

Table 45: Gender related transportation issues

Variables	Perceptions	Male		Female	
		Count	Column N %	Count	Column N %
"Traffic congestion that you experience while driving"	Not a Problem	403	32.0%	265	27.5%
	Problem	607	48.2%	520	54.0%
	Major Problem	249	19.8%	178	18.5%
"Traffic noise that you hear at home, work or school"	Not a Problem	136	10.8%	84	8.7%
	Problem	795	62.9%	626	64.9%
	Major Problem	332	26.3%	255	26.4%
"Vehicle emissions that affect local air quality"	Not a Problem	265	21.0%	150	15.6%
	Problem	551	43.7%	473	49.2%
	Major Problem	444	35.2%	338	35.2%
"Vehicle emissions that contribute to climate change"	Not a Problem	126	10.0%	91	9.5%
	Problem	660	52.4%	479	49.8%
	Major Problem	474	37.6%	391	40.7%
"Unsafe communities because of speeding traffic"	Not a Problem	264	21.0%	140	14.6%
	Problem	600	47.8%	476	49.6%
	Major Problem	391	31.2%	343	35.8%
"Importing much of our oil from foreign countries"	Not a Problem	125	9.9%	107	11.2%
	Problem	602	47.9%	452	47.2%
	Major Problem	531	42.2%	399	41.6%
Cells highlighted represent the highest percentage across gender					

3.4.8 Impact on environment

Table 46 explains the impact of the transport sector on the environment and health. Both male and female respondents agree with the statement that "cars, minivans, vans, pickups and SUVs are not an important source of air pollution anymore" because of the changed technology. However, male respondents agreed with the statement that "cars, minivans, vans, pickups and SUVs are an important source of the greenhouse gases that many scientists believe are warming the earth's climate". For the rest of the variables, no clear majority was observed in the analysis.

Table 46: Environmental issues by gender

Variables	Perceptions	Male		Female	
		Count	Column N %	Count	Column N %
"Cars, minivans, vans, pickups and SUVs are not an important source of air pollution anymore"	Agree	780	63.0%	590	61.5%
	Disagree	368	29.7%	321	33.5%
	Unsure	91	7.3%	48	5.0%
"Government rules allow minivans, vans, pickups and SUVs to pollute more than passenger cars, for every gallon of gas used"	Agree	502	40.7%	424	44.4%
	Disagree	547	44.4%	363	38.0%
	Unsure	184	14.9%	169	17.7%
"Cars, minivans, vans, pickups and SUVs are an important source of the greenhouse gases that many scientists believe are warming the earth's climate"	Agree	654	53.0%	464	48.3%
	Disagree	371	30.1%	274	28.5%
	Unsure	209	16.9%	222	23.1%
"Government rules require minivans, vans, pickups and SUVs to meet the same miles-per-gallon standards as passenger cars"	Agree	436	35.3%	322	33.7%
	Disagree	511	41.4%	371	38.8%
	Unsure	288	23.3%	263	27.5%
"Exhaust from cars, minivans, vans, pickups and SUVs is an important source of the pollution that causes asthma and makes asthma attacks worse"	Agree	594	47.9%	408	42.5%
	Disagree	375	30.3%	307	32.0%
	Unsure	270	21.8%	244	25.4%
Cells highlighted represent the highest percentage across genders					

3.4.9 Environment related gender preferences

Table 47 explains that for both male and female respondents "reduced impact on the environment" is an important variable for them and it is important for them to drive a vehicle with more advanced or innovative technology.

Table 47: Environment related gender preferences

Variables	Perceptions	Male		Female	
		Count	Column N %	Count	Column N %
Saving money on the cost of operation (using electricity rather than gasoline)	Not Important	521	41.7%	371	38.6%
	Important	510	40.8%	459	47.8%
	Very Important	219	17.5%	130	13.5%
Reduced impact on the environment	Not Important	107	8.5%	76	7.9%
	Important	805	64.3%	621	64.8%
	Very Important	340	27.2%	261	27.2%
Reduced dependence on gasoline	Not Important	331	26.5%	213	22.2%
	Important	532	42.5%	452	47.1%
	Very Important	388	31.0%	294	30.7%
Driving a vehicle with more advanced or innovative technology	Not Important	117	9.4%	94	9.8%
	Important	716	57.3%	539	56.1%
	Very Important	416	33.3%	328	34.1%
Cells highlighted represent the highest percentage across genders					

Table 48 highlights that "higher purchase price for EV than for a comparable conventional vehicle" is an important factor for both male and female respondents and also that the need to plug the vehicle in to recharge the battery is an important factor for them. Female respondents were concerned about the vehicle size and considered it an important factor. For the rest of the variables, no clear majority was observed in the analysis.

Table 48: Gender wise preferences for important factors

Variables	Perceptions	Male		Female	
		Count	Column N %	Count	Column N %
A higher purchase price than for a comparable conventional vehicle	Not Important	316	25.5%	249	26.0%
	Important	702	56.7%	542	56.6%
	Very Important	221	17.8%	166	17.3%
The need to plug the vehicle into recharge the battery	Not Important	116	9.3%	83	8.7%
	Important	715	57.6%	608	63.5%
	Very Important	411	33.1%	267	27.9%
Concerns about limited access to plug-in locations	Not Important	196	15.9%	122	12.8%
	Important	598	48.4%	461	48.3%
	Very Important	442	35.8%	372	39.0%
Availability of desirable vehicle size or style	Not Important	148	11.9%	102	10.7%
	Important	591	47.6%	494	51.7%
	Very Important	502	40.5%	360	37.7%
Reliability of the vehicle	Not Important	181	14.6%	123	12.9%
	Important	557	45.0%	439	46.0%
	Very Important	501	40.4%	393	41.2%
On-going maintenance and operative costs (including battery replacement)	Not Important	82	6.6%	47	4.9%
	Important	573	46.3%	461	48.2%
	Very Important	583	47.1%	449	46.9%
The ability to carry heavy loads	Not Important	206	16.6%	117	12.2%
	Important	525	42.3%	423	44.1%
	Very Important	510	41.1%	419	43.7%
Cells highlighted represent the highest percentage across gender					

3.4.10 Factors important for the purchase of personal vehicles based on gender

Table 49 highlights that fuel efficiency, safety, vehicle power and reliability are very important factors for both male and female respondents. Purchase price, vehicle size, vehicle reputation, vehicle emission and pollution are important factors for both male and females. Males were concerned about fuel type and consider this very important.

Table 49: Factors important for purchase of personal vehicles by gender

Variables	Perceptions	Male		Female	
		Count	%	Count	%
Fuel efficiency	Very Important	1053	83.1%	792	82.0%
	Important	197	15.5%	165	17.1%
	Not Important	17	1.3%	9	.9%
Safety	Very Important	755	59.5%	588	60.9%
	Important	493	38.9%	368	38.1%
	Not Important	20	1.6%	10	1.0%
Vehicle power	Very Important	800	63.2%	615	63.7%
	Important	404	31.9%	309	32.0%
	Not Important	61	4.8%	41	4.2%
Purchase price	Very Important	532	42.1%	435	45.0%
	Important	689	54.5%	501	51.8%
	Not Important	44	3.5%	31	3.2%
Reliability	Very Important	808	64.0%	561	58.5%
	Important	403	31.9%	361	37.6%
	Not Important	52	4.1%	37	3.9%
Vehicle size	Very Important	373	29.6%	316	32.8%
	Important	798	63.3%	575	59.7%
	Not Important	90	7.1%	72	7.5%
Expected operating costs	Very Important	626	49.6%	428	44.4%
	Important	560	44.4%	476	49.3%
	Not Important	76	6.0%	61	6.3%
Reputation of particular vehicle make or model	Very Important	375	29.9%	289	30.2%
	Important	771	61.5%	588	61.4%
	Not Important	107	8.5%	80	8.4%
Fuel type (e.g. petrol, diesel, CNG)	Very Important	644	51.0%	439	45.6%
	Important	510	40.4%	447	46.4%
	Not Important	109	8.6%	77	8.0%
Vehicle emissions and pollution	Very Important	468	37.1%	358	37.1%
	Important	702	55.6%	532	55.1%
	Not Important	92	7.3%	76	7.9%
Cells highlighted represent the highest percentage across gender					

3.4.11 Premium for electric vehicle

Both male and female respondents were willing to pay a premium for an electric vehicle. It was revealed that respondents were ready to pay 15% premium over the price of a comparable conventional vehicle to save the environment (Table 50).

Table 50: Willing to pay a premium for electric vehicle

	Premium	Gender			
		Male		Female	
		Count	Column N %	Count	Column N %
If very/somewhat interested in electric vehicles would you be willing to pay a premium to purchase an electric vehicle?	10%	365	30.4%	256	27.5%
	15%	503	41.8%	421	45.2%
	20%	334	27.8%	254	27.3%
Cells highlighted represent the highest percentage across gender					

3.5 Preferences based on education level

To understand the perceptions based on education level, we asked multiple questions to the respondents. The questions were asked to know the average daily travel, different types of information sources, who influences their purchasing decisions, their views on purchasing electric vehicles, what they think about the role of State and Central Governments, important factors for purchasing any vehicle, different types of transport issues, various reasons which can influence their decisions to purchase an electric vehicle in the future, impact of transport sector on the environment, especially vehicular pollution and finally the perception on the premium for electric vehicles. The results were analyzed based on education level and are summarized below.

3.5.1 Estimated average daily drive

Table 51 highlights that the percentage of students who travel less than 10 km in a day is higher than compared to those respondents who had completed their bachelor’s and master’s degrees. The respondents who had completed their bachelor's and master's degrees travel upto 20 km ina day to reach their offices. From thisdata, it can be concluded that most of the respondents from all age groups travel upto 60 km in a day.

Table 51: Estimated average daily travel

		Education					
		Bachelor's		Master's		Student	
		Count	Column N %	Count	Column N %	Count	Column N %
Drive per day	Up to10 km	190	19.0%	78	9.0%	75	28.8%
	11 to 20 km	349	34.9%	256	29.7%	62	23.8%
	21 to 30 km	197	19.7%	193	22.4%	35	13.5%
	31-60 km	188	18.8%	245	28.4%	55	21.2%
	60 km or more	75	7.5%	91	10.5%	33	12.7%
Cells highlighted represent the highest percentage across education level							

3.5.2 Information sources

Table 52 highlights that respondents with different education backgrounds had different sources of information. Television was the main source of information for those respondents who had completed their bachelor's degree. For respondents who had completed their master's degree, radio was the main source of information. However, students were mostly dependent on word of mouth and other information sources for knowing the latest developments around them.

Table 52: Information source for new technologies

	Education					
	Bachelor's		Master's		Student	
	Count	Row N %	Count	Row N %	Count	Row N %
Television	689	52.7%	488	37.3%	131	10.0%
Newspaper	651	49.1%	557	42.0%	119	9.0%
Radio	182	47.9%	188	49.5%	10	2.6%
Magazines	394	49.7%	326	41.1%	73	9.2%
Internet	614	44.1%	598	43.0%	180	12.9%
Word of mouth	120	50.4%	81	34.0%	37	15.5%
Other	85	44.0%	78	40.4%	30	15.5%
Cells highlighted represent the highest percentage across education level						

3.5.3 Influence on purchasing decision

Table 53 revealed that elders influence the purchasing decisions of all respondents irrespective of education levels. For the majority of the respondents, elders take the decision for any vehicle purchase. However students shared that the decision is taken jointly.

Table 53: Vehicle purchase decision – influence source

		Education					
		Bachelor's		Master's		Student	
		Count	Column N %	Count	Column N %	Count	Column N %
Who influences your vehicle purchase decision?	Elders	615	59.4%	414	47.4%	126	44.2%
	Children	109	10.5%	144	16.5%	33	11.6%
	Joint decision	311	30.0%	315	36.1%	126	44.2%

Cells highlighted represent the highest percentage across education level

3.5.4 About new vehicles

Table 54 highlights that respondents from all the age groups wait to read reviews and then buy the vehicle only if the reviews are favorable. Students wait until the new technology has been widely accepted and proven before considering it.

Table 54: When a new vehicle becomes available for purchase, what do you do?

		Education					
		Bachelor's		Master's		Student	
		Count	Column N %	Count	Column N %	Count	Column N %
When a new vehicle becomes available for purchase, what do you do?	I am among the first to purchase it.	170	16.5%	113	12.9%	38	12.5%
	I wait to read a review of it and then buy it if the review is favorable.	483	46.9%	369	42.3%	124	40.8%
	I wait until this new technology has been widely accepted and proven before considering it.	354	34.4%	363	41.6%	124	40.8%
	Other.	22	2.1%	28	3.2%	18	5.9%

Cells highlighted represent the highest percentage across education level

3.5.5 Interested in new electric vehicles

Respondents were ready to purchase electric vehicles once they become easily available in the next couple of years. Majority of respondents from all education groups were "somewhat interested" in purchasing an electric motor vehicle once they become easily available in the next couple of years (Table 55).

Table 55: Interest level in purchasing electric vehicle

		Education					
		Bachelor's		Master's		Student	
		Count	Column N %	Count	Column N %	Count	Column N %
How interested would you be in purchasing an electric motor vehicle once they become easily available in the next couple of years?	Very interested	335	32.1%	224	25.7%	82	27.1%
	Somewhat interested	405	38.8%	365	41.9%	128	42.2%
	Not very interested	128	12.3%	125	14.4%	32	10.6%
	Not planning to purchase future vehicles	96	9.2%	82	9.4%	29	9.6%
	Cannot say	79	7.6%	75	8.6%	32	10.6%

Cells highlighted represent the highest percentage across education level

3.5.6 Role of State and Central Government

Table 56 revealed that respondents from all education groups felt that the State and Central Governments should encourage electric vehicles for both public and private transport by making amendments in the transport policy. Only 9% of the respondents who were students shared that the policies defined by State and Central Government are not worth promoting, for both the sectors.

Table 56: Encouragement of electric vehicles by State and Central Government

		Education					
		Bachelor's		Master's		Student	
		Count	Column N %	Count	Column N %	Count	Column N %
Do you think transport policy of the State and Central Government should encourage electric vehicles?	Yes, but only for public transport	341	33.0%	278	32.1%	73	24.1%
	Yes, both for public and private transport	546	52.8%	472	54.4%	202	66.7%
	No, not worth it	147	14.2%	117	13.5%	28	9.2%
Cells highlighted represent the highest percentage across education level							

3.5.7 Factors important for purchase of personal vehicles – education wise

For all the education groups, fuel efficiency, safety, vehicle power and reliability were very important factors. Vehicle purchase price and vehicle size were important factors for all respondents from all age groups. Operating cost of the vehicle was a very important factor for those respondents who had completed their bachelor's and master's degrees, however for students it was just an important factor. Students were more concerned about vehicle emissions and pollution and considered this a very important factor while other groups of respondents considered it an important factor for them (Table 57).

Table 57: Important factors for purchasing personal vehicles by region

		Education					
		Bachelor's		Master's		Student	
		Count	Column N %	Count	Column N %	Count	Column N %
Fuel efficiency	Very Important	881	83.8%	725	82.4%	248	80.5%
	Important	156	14.8%	145	16.5%	58	18.8%
	Not Important	14	1.3%	10	1.1%	2	.6%
Safety	Very Important	628	59.8%	477	54.2%	243	78.4%
	Important	407	38.8%	389	44.2%	66	21.3%
	Not Important	15	1.4%	14	1.6%	1	.3%
Vehicle power	Very Important	714	68.1%	550	62.5%	164	53.2%
	Important	301	28.7%	273	31.0%	130	42.2%
	Not Important	34	3.2%	57	6.5%	14	4.5%
Purchase price	Very Important	456	43.4%	370	42.1%	145	47.1%
	Important	561	53.4%	478	54.4%	154	50.0%
	Not Important	33	3.1%	31	3.5%	9	2.9%
Reliability	Very Important	647	61.7%	542	61.6%	187	61.7%
	Important	361	34.4%	297	33.8%	108	35.6%
	Not Important	40	3.8%	41	4.7%	8	2.6%
Vehicle size (to accommodate passengers or cargo)	Very Important	306	29.3%	291	33.1%	94	30.7%
	Important	680	65.1%	520	59.1%	180	58.8%
	Not Important	59	5.6%	69	7.8%	32	10.5%
Expected operating costs (for maintenance and repair)	Very Important	504	48.1%	436	49.5%	125	40.8%
	Important	489	46.7%	382	43.4%	159	52.0%
	Not Important	54	5.2%	62	7.0%	22	7.2%
Reputation of particular vehicle make or model	Very Important	311	29.8%	263	29.9%	97	32.6%
	Important	655	62.8%	543	61.8%	164	55.0%
	Not Important	77	7.4%	73	8.3%	37	12.4%
Fuel type (e.g. petrol, diesel, CNG)	Very Important	533	50.8%	414	47.0%	144	47.5%
	Important	424	40.4%	393	44.7%	136	44.9%
	Not Important	93	8.9%	73	8.3%	23	7.6%
Vehicle emissions and pollution	Very Important	381	36.4%	276	31.3%	173	56.4%
	Important	587	56.1%	534	60.6%	115	37.5%
	Not Important	78	7.5%	71	8.1%	19	6.2%
Cells highlighted represent the highest percentage across education level							

3.5.8 Transport issues

Respondents from all education groups shared that various factors such as traffic congestion while driving, speeding traffic and traffic noise at home, work or school, are a problem for them and has created a lot of issues in society. The other factors such as vehicle emissions that affect local air quality and importing fuel from foreign countries were major problems for students, however for the other education groups, it was just a problem (Table 58).

Table 58: Transport issues based on education levels

		Education					
		Bachelor's		Master's		Student	
		Count	Column N %	Count	Column N %	Count	Column N %
Traffic congestion that you experience while driving.	Not a Problem	366	35.0%	274	31.1%	32	10.6%
	Problem	512	48.9%	473	53.8%	141	46.7%
	Major Problem	169	16.1%	133	15.1%	129	42.7%
Traffic noise that you hear at home, work or school	Not a Problem	119	11.4%	80	9.1%	24	7.8%
	Problem	680	64.9%	562	63.9%	176	57.5%
	Major Problem	249	23.8%	238	27.0%	106	34.6%
Vehicle emissions that affect local air quality	Not a Problem	247	23.6%	159	18.1%	11	3.6%
	Problem	462	44.2%	432	49.2%	131	43.1%
	Major Problem	336	32.2%	287	32.7%	162	53.3%
Vehicle emissions that contribute to climate change	Not a Problem	100	9.6%	101	11.5%	18	5.9%
	Problem	558	53.4%	460	52.3%	127	41.8%
	Major Problem	386	37.0%	319	36.3%	159	52.3%
Unsafe communities because of speeding traffic	Not a Problem	221	21.2%	163	18.5%	19	6.3%
	Problem	488	46.8%	442	50.2%	155	51.5%
	Major Problem	334	32.0%	275	31.3%	127	42.2%
Importing much of our oil from foreign countries	Not a Problem	103	9.9%	93	10.6%	38	12.6%
	Problem	516	49.4%	407	46.3%	130	43.0%
	Major Problem	425	40.7%	379	43.1%	134	44.4%
Cells highlighted represent the highest percentage across education level							

3.5.9 Various reasons for purchasing electric vehicles in the future

Table 59 highlights that saving money on the cost of operations was an important factor for those respondents who have completed their bachelor's degree or are studying currently, but it was not an important factor for those respondents who have completed their master's degree.

Respondents from all the education groups shared that it is important to drive a vehicle with more advanced or innovative technology and reduce the impact on the environment by using electric vehicles which will also reduce dependence on gasoline.

Table 59: Education-wise preferences for various attributes

		Education					
		Bachelor's		Master's		Student	
		Count	Column N %	Count	Column N %	Count	Column N %
Saving money on the cost of operation (using electricity rather than gasoline)	Not Important	444	42.7%	414	47.5%	39	12.9%
	Important	453	43.6%	341	39.1%	177	58.4%
	Very Important	143	13.8%	117	13.4%	87	28.7%
Reduced impact on the environment	Not Important	95	9.1%	69	7.9%	20	6.6%
	Important	674	64.6%	608	70.0%	148	49.0%
	Very Important	275	26.3%	192	22.1%	134	44.4%
Reduced dependence on gasoline	Not Important	293	28.1%	224	25.8%	25	8.3%
	Important	444	42.6%	377	43.4%	165	54.5%
	Very Important	306	29.3%	268	30.8%	113	37.3%
Driving a vehicle with more advanced or innovative technology	Not Important	105	10.1%	83	9.6%	27	8.9%
	Important	560	53.7%	531	61.1%	169	55.8%
	Very Important	378	36.2%	255	29.3%	107	35.3%
Cells highlighted represent the highest percentage across education levels							

Variables such as, "higher purchase price than for a comparable conventional vehicle, the need to plug the vehicle in to recharge the battery, concerns about limited access to plug-in locations, availability of desirable vehicle size or style, on-going maintenance & operative costs and the ability to carry heavy loads" are important factors for all the respondents from all education levels. Student respondents were more concerned about vehicle reliability and shared that it is a very important factor for them, however the other education groups shared that it is just an important factor (Table 60).

Table 60: Factors supporting purchase of electric vehicles by education level

		Education					
		Bachelor's		Master's		Student	
		Count	Column N %	Count	Column N %	Count	Column N %
A higher purchase price than for a comparable conventional vehicle	Not Important	282	27.2%	243	27.9%	41	13.8%
	Important	599	57.8%	462	53.1%	188	63.3%
	Very Important	156	15.0%	165	19.0%	68	22.9%
The need to plug the vehicle in to recharge the battery	Not Important	95	9.1%	66	7.6%	36	12.2%
	Important	639	61.4%	532	61.1%	163	55.1%
	Very Important	306	29.4%	272	31.3%	97	32.8%
Concerns about limited access to plug-in locations	Not Important	169	16.3%	117	13.4%	32	11.0%
	Important	520	50.2%	404	46.4%	135	46.2%
	Very Important	347	33.5%	349	40.1%	125	42.8%
Availability of desirable vehicle size or style	Not Important	118	11.4%	90	10.4%	44	14.9%
	Important	495	47.7%	436	50.2%	161	54.4%
	Very Important	425	40.9%	342	39.4%	91	30.7%
Reliability of the vehicle	Not Important	148	14.3%	133	15.3%	22	7.5%
	Important	475	45.8%	397	45.7%	127	43.2%
	Very Important	414	39.9%	339	39.0%	145	49.3%
On-going maintenance and operative costs (including battery replacement)	Not Important	59	5.7%	48	5.5%	19	6.5%
	Important	481	46.3%	422	48.6%	142	48.3%
	Very Important	499	48.0%	399	45.9%	133	45.2%
The ability to carry heavy loads	Not Important	180	17.3%	112	12.9%	32	10.8%
	Important	438	42.2%	381	43.9%	134	45.1%
	Very Important	421	40.5%	375	43.2%	131	44.1%
Cells highlighted represent the highest percentage across education level							

3.5.10 Impact on environment

Table 61 explains that respondents who have completed their bachelor's and master’s degrees agreed with the statements that “cars, minivans, vans, pickups and SUVs are not an important source of air pollution anymore”, but the respondents who were students disagreed with this statement. Respondents from all education levels agreed with the statement that “cars, minivans, vans, pickups and SUVs are an important source of the greenhouse gases that many scientists believe are warming the earth’s climate”.

Respondents who have completed their bachelor's and master’s degrees agreed with the statements that “cars, minivans, vans, pickups and SUVs are an important source of greenhouse gases that many scientists believe are warming the earth's climate", however student respondents disagreed with this statement.

Table 61: Breakup of impact on environment of various transport modes by education level

		Education					
		Bachelor's		Master's		Student	
		Count	Column N %	Count	Column N %	Count	Column N %
Cars, minivans, vans, pickups and SUVs are not an important source of air pollution anymore.	Agree	707	68.1%	546	62.4%	121	41.4%
	Disagree	272	26.2%	272	31.1%	147	50.3%
	Unsure	59	5.7%	57	6.5%	24	8.2%
Government rules allow minivans, vans, pickups and SUVs to pollute more than passenger cars, for every gallon of gas used.	Agree	459	44.2%	333	38.3%	131	45.3%
	Disagree	442	42.6%	378	43.4%	93	32.2%
	Unsure	137	13.2%	159	18.3%	65	22.5%
Cars, minivans, vans, pickups and SUVs are an important source of the greenhouse gases that many scientists believe are warming the earth's climate	Agree	562	54.2%	365	41.8%	190	65.5%
	Disagree	295	28.4%	299	34.2%	53	18.3%
	Unsure	180	17.4%	210	24.0%	47	16.2%
Government rules require minivans, vans, pickups and SUVs to meet the same miles-per-gallon standards as passenger cars.	Agree	374	36.1%	258	29.6%	126	43.4%
	Disagree	444	42.9%	356	40.8%	87	30.0%
	Unsure	218	21.0%	259	29.7%	77	26.6%
Exhaust from cars, minivans, vans, pickups and SUVs is an important source of the pollution that causes asthma and makes asthma attacks worse	Agree	509	49.1%	315	36.0%	182	62.3%
	Disagree	300	28.9%	320	36.5%	60	20.5%
	Unsure	228	22.0%	241	27.5%	50	17.1%
Cells highlighted represent the highest percentage across education level							

3.5.11 Premium for electric vehicles

All the respondents from all the education groups were willing to pay a premium ranging from 10% to 20%. The majority of the respondents from all the education groups were willing to pay 15% premium for electric vehicles (Table 62).

Table 62: Willing to pay a premium for electric vehicle

		Education					
		Bachelor's		Master's		Student	
		Count	Column N %	Count	Column N %	Count	Column N %
If very/somewhat interested in electric vehicles would you be willing to pay a premium to purchase an electric vehicle?	10%	327	32.4%	227	26.4%	77	27.5%
	15%	397	39.3%	408	47.4%	120	42.9%
	20%	285	28.2%	225	26.2%	83	29.6%
Cells highlighted represent the highest percentage across education levels							

3.6 Preferences based on income groups

To understand the perceptions based on income groups, we asked multiple questions to the respondents. The questions were asked to know the average daily travel, different types of information sources, who influences their purchasing decisions, their views on purchasing electric vehicles, what they think about the role of State and Central Governments, important factors for purchasing any vehicle, different types of transport issues, various reasons which can influence their decisions to purchase an electric vehicle in future, impact of transport sector on environment, especially vehicular pollution and finally their perceptions on paying a premium for electric vehicles. The results were analyzed by monthly income and are summarized below.

3.6.1 Estimated average daily drive

Table 63 highlights that respondents with higher incomes can afford to travel more to finish their work. Approximately 82 percent of respondents who fall under the monthly income bracket of INR 25,000-50,000 travel upto 30 km. Similarly 59% of the respondents who fall under the monthly income bracket of INR 50,000-100,000 travel upto 30 km and 51% of the respondents who have incomes higher than INR 1 lakh per month undertake travel upto 30 km in a day.

Table 63: Estimated average daily travel

		Monthly income (INR)					
		25,000 – 50,000		50,000 – 100,000		Above 1 lakh	
		Count	Column N %	Count	Column N %	Count	Column N %
Drive per day	Up to10 km	206	17.8%	66	9.6%	22	16.5%
	11 to 20 km	412	35.6%	194	28.2%	26	19.5%
	21 to 30 km	229	19.8%	147	21.4%	20	15.0%
	31-60 km	239	20.7%	186	27.0%	35	26.3%
	60 km or more	70	6.1%	95	13.8%	30	22.6%
Cells highlighted represent the highest percentage across income groups							

3.6.2 Information sources

Table 64 reveals that there is a difference between sources of information among monthly income groups. In income group of INR 25,000-50,000 television is the main source of information followed by newspapers and the internet. For the income groups of INR 50,000-100,000 and above 1 lakh, the internet remains the prime source of information, followed by newspapers, television, magazines and radio.

Table 64: Information sources for new technologies

	Monthly income (INR)					
	25,000 – 50,000		50,000 – 100,000		Above 1 lakh	
	Count	Column N %	Count	Column N %	Count	Column N %
Television	770	63.7%	369	52.8%	54	39.4%
Newspaper	749	62.0%	431	61.7%	65	47.4%
Radio	200	16.6%	147	21.0%	23	16.8%
Magazines	442	36.6%	262	37.5%	44	32.1%
Internet	736	61.0%	477	68.2%	93	67.9%
Word of mouth	113	9.4%	72	10.3%	22	16.1%
Other	102	8.4%	66	9.5%	11	8.0%
Cells highlighted represent the highest percentage across income groups						

3.6.3 Influence on purchasing decision

Data in Table 65 shows that respondents from income group of INR 25,000-50,000 and INR 50,000-100,000 consult with elders while taking their decision about purchasing any vehicle. The respondents who had a monthly income of more than INR1 lakh per month were taking a joint decision about purchasing any vehicle.

Table 65: Vehicle purchase decision – influence source

		Monthly income (INR)					
		25,000 – 50,000		50,000 – 100,000		Above 1 lakh	
		Count	Column N %	Count	Column N %	Count	Column N %
Who influences your vehicle purchase decision?	Elders	710	59.7%	330	47.3%	44	32.1%
	Children	147	12.4%	110	15.8%	27	19.7%
	Joint decision	333	28.0%	257	36.9%	66	48.2%
Cells highlighted represent the highest percentage across income groups							

3.6.4 About new vehicle

Table 66 highlights that in all the income groups, respondents shared that they would wait to read the reviews and then buy it if the reviews are favorable. Respondents also wanted to wait until the new technology is widely accepted and proven in the market. Data also revealed that people from the income group of INR 25,000-50,000 are ready to purchase a new vehicle compared to respondents from other income groups.

Table 66: When a new vehicle becomes available for purchase, what do you do?

		Monthly income (INR)					
		25,000 – 50,000		50,000 – 100,000		Above 1 lakh	
		Count	Column N %	Count	Column N %	Count	Column N %
When a new vehicle becomes available for purchase, what do you do?	I am among the first to purchase it.	218	18.2%	85	12.2%	13	9.6%
	I wait to read a review of it and then buy it if the review is favorable.	540	45.2%	296	42.5%	57	42.2%
	I wait until this new technology has been widely accepted and proven before considering it.	408	34.1%	292	42.0%	54	40.0%
	Other.	29	2.4%	23	3.3%	11	8.1%
Cells highlighted represent the highest percentage across income groups							

3.6.5 Interested in new electric vehicles

Table 67 revealed that respondents from all income groups were very interested in purchasing an electric motor vehicle once they become easily available in the next couple of years but they were "somewhat interested" in purchasing such vehicles. There is no significant difference which exists between the respondents who are not planning to purchase an electric vehicle in all the income groups.

Table 67: Interest level in purchasing electric vehicle

		Monthly income (INR)					
		25,000 – 50,000		**50,000 – 100,000**		**Above 1 lakh**	
		Count	**Column N %**	**Count**	**Column N %**	**Count**	**Column N %**
How interested would you be in purchasing an electric motor vehicle once they become easily available in the next couple of years?	Very interested	375	31.2%	161	23.2%	43	32.1%
	Somewhat interested	487	40.5%	290	41.7%	49	36.6%
	Not very interested	140	11.6%	110	15.8%	18	13.4%
	Not planning to purchase future vehicles	120	10.0%	67	9.6%	11	8.2%
	Cannot say	80	6.7%	67	9.6%	13	9.7%
Cells highlighted represent the highest percentage across income groups							

3.6.6 Role of State and Central Government

Table 68 reveals that respondents from all income groups shared that the State and Central Government should encourage electric vehicles for public transport. Respondents also shared that the State and Central Government should encourage electric vehicles for private transport.

Table 68: Encouragement of electric vehicles by State and Central Government

		Monthly income (INR)					
		25,000 – 50,000		50,000 – 100,000		Above 1 lakh	
		Count	Column N %	Count	Column N %	Count	Column N %
Do you think transport policy of the State and Central Government should encourage electric vehicles?	Yes, but only for public transport	403	34.0%	211	30.4%	45	33.6%
	Yes, both for public and private transport	622	52.4%	382	55.0%	68	50.7%
	No, not worth it	162	13.6%	102	14.7%	21	15.7%
Cells highlighted represent the highest percentage across income groups							

3.6.7 Factors important for purchase of personal vehicles – income-wise

Table 69 shows that factors such as fuel efficiency, safety, vehicle power and reliability of the vehicle were very important for purchasing of a personal vehicle, across income groups. Vehicle size, emissions and pollution are important factors for all the income groups. There is a difference between the perceptions of respondents on fuel type and operating cost of the vehicles. For the monthly income group of INR 25,000-50,000 and above INR 1 lakh, they are very important factors, however for the income group of INR 50,000-100,000 it is just an important factor.

Table 69: Important factors for purchasing of personal vehicles by income group

		Monthly income (INR)					
		25,000 – 50,000		50,000 – 100,000		Above 1 lakh	
		Count	Column N %	Count	Column N %	Count	Column N %
Fuel efficiency	Very Important	1045	86.3%	567	81.0%	96	70.6%
	Important	155	12.8%	123	17.6%	37	27.2%
	Not Important	11	.9%	10	1.4%	3	2.2%
Safety	Very Important	734	60.6%	375	53.6%	73	53.3%
	Important	463	38.2%	313	44.7%	61	44.5%
	Not Important	14	1.2%	12	1.7%	3	2.2%
Vehicle power	Very Important	820	67.8%	420	60.3%	81	59.1%
	Important	355	29.3%	218	31.3%	49	35.8%
	Not Important	35	2.9%	59	8.5%	7	5.1%
Purchase price	Very Important	523	43.2%	307	44.0%	50	36.5%
	Important	657	54.3%	362	51.9%	83	60.6%
	Not Important	31	2.6%	29	4.2%	4	2.9%
Reliability	Very Important	772	64.0%	406	57.9%	83	61.0%
	Important	392	32.5%	261	37.2%	45	33.1%
	Not Important	43	3.6%	34	4.9%	8	5.9%
Vehicle size (to accommodate passengers or cargo)	Very Important	350	29.1%	237	33.8%	47	34.6%
	Important	785	65.3%	399	56.9%	82	60.3%
	Not Important	68	5.7%	65	9.3%	7	5.1%
Expected operating costs (for maintenance and repair)	Very Important	583	48.3%	323	46.1%	74	54.4%
	Important	553	45.8%	326	46.6%	54	39.7%
	Not Important	72	6.0%	51	7.3%	8	5.9%
Reputation of particular vehicle make or model	Very Important	332	27.8%	211	30.3%	48	35.3%
	Important	764	63.9%	426	61.1%	82	60.3%
	Not Important	100	8.4%	60	8.6%	6	4.4%
Fuel type (e.g. petrol, diesel, CNG)	Very Important	603	50.0%	307	43.9%	76	55.5%
	Important	500	41.5%	330	47.1%	55	40.1%
	Not Important	102	8.5%	63	9.0%	6	4.4%
Vehicle emissions and pollution	Very Important	414	34.2%	240	34.3%	48	35.0%
	Important	697	57.6%	403	57.6%	82	59.9%
	Not Important	99	8.2%	57	8.1%	7	5.1%
Cells highlighted represent the highest percentage across income groups							

3.6.8 Transport issues

For a better understanding of the issues related to the transport system, questions focusing on the problems faced by the respondents were asked. Various variables were included such as "congestion and noise; vehicle emissions and pollution; traffic speed; global warming; fuel import" etc. These variables highlighted the preferences of respondents while buying any vehicle. In Table 41, we have explained the various issues based on different income groups. The majority of the respondents expressed that these factors are a problem for them (Table 70).

Table 70: Transport issues based on income group

		Monthly income (INR)					
		25,000 – 50,000		50,000 – 100,000		Above 1 lakh	
		Count	Column N %	Count	Column N %	Count	Column N %
Traffic congestion that you experience while driving.	Not a Problem	401	33.2%	224	32.0%	31	22.6%
	Problem	613	50.8%	365	52.2%	67	48.9%
	Major Problem	193	16.0%	110	15.7%	39	28.5%
Traffic noise that you hear at home, work or school	Not a Problem	135	11.2%	64	9.2%	13	9.5%
	Problem	772	63.8%	463	66.2%	79	57.7%
	Major Problem	303	25.0%	172	24.6%	45	32.8%
Vehicle emissions that affect local air quality	Not a Problem	261	21.6%	127	18.2%	24	17.5%
	Problem	570	47.2%	325	46.5%	61	44.5%
	Major Problem	377	31.2%	247	35.3%	52	38.0%
Vehicle emissions that contribute to climate change	Not a Problem	113	9.4%	88	12.6%	15	10.9%
	Problem	630	52.2%	373	53.4%	69	50.4%
	Major Problem	465	38.5%	238	34.0%	53	38.7%
Unsafe communities because of speeding traffic	Not a Problem	238	19.8%	129	18.5%	25	18.4%
	Problem	581	48.2%	342	49.0%	59	43.4%
	Major Problem	386	32.0%	227	32.5%	52	38.2%
Importing much of our oil from foreign countries	Not a Problem	109	9.1%	84	12.0%	16	11.7%
	Problem	587	48.8%	314	45.0%	64	46.7%
	Major Problem	508	42.2%	300	43.0%	57	41.6%

Cells highlighted represent the highest percentage across income groups

3.6.9 Various reasons for purchasing electric vehicles in the future

Table 71 reveals that respondents from the monthly income group INR 25,000-50,000 are unable to decide whether or not this is an important factor: "saving money on the cost of operations (using electricity rather than gasoline)". Respondents from the income group of INR 50,000-100,000 shared that this is not an important factor for them, however the respondents who had monthly income of more than INR 1 lakh considered that an important factor while purchasing any vehicle.

Respondents from all the income groups shared that it is important to drive a vehicle with more advanced or innovative technology and reduce the impact on the environment by using electric vehicles which will also reduce dependence on gasoline.

Table 71: Preferences for various attributes based on income groups

		Monthly income (INR)					
		25,000 – 50,000		50,000 – 100,000		Above 1 lakh	
		Count	Column N %	Count	Column N %	Count	Column N %
Saving money on the cost of operation (using electricity rather than gasoline)	Not Important	519	43.2%	323	46.3%	49	36.3%
	Important	519	43.2%	287	41.2%	56	41.5%
	Very Important	164	13.6%	87	12.5%	30	22.2%
Reduced impact on the environment	Not Important	116	9.7%	54	7.7%	11	8.1%
	Important	771	64.3%	501	71.8%	86	63.7%
	Very Important	313	26.1%	143	20.5%	38	28.1%
Reduced dependence on gasoline	Not Important	319	26.6%	165	23.6%	40	29.6%
	Important	562	46.9%	287	41.1%	48	35.6%
	Very Important	318	26.5%	246	35.2%	47	34.8%
Driving a vehicle with more advanced or innovative technology	Not Important	120	10.0%	62	8.9%	18	13.3%
	Important	672	56.0%	425	60.8%	76	56.3%
	Very Important	407	33.9%	212	30.3%	41	30.4%
Cells highlighted represent the highest percentage across income groups							

Respondents across income groups shared that variables such as higher purchase price than for a comparable conventional vehicle, recharging facilities and availability of desirable vehicle size or style are important factors. Respondents from the income group of INR 25,000-50,000 and INR 50,000-100,000 were concerned about there liability of the vehicle, on-going maintenance and operative costs, and ability to carry heavy loads, but for the respondents who had an income of more than INR1 lakh these factors were very important (Table 72).

Table 72: Factors supporting purchase of electric vehicles by income group

		Monthly income (INR)					
		25,000 – 50,000		50,000 – 100,000		Above 1 lakh	
		Count	Column N %	Count	Column N %	Count	Column N %
A higher purchase price than for a comparable conventional vehicle	Not Important	309	25.9%	201	28.9%	29	21.6%
	Important	700	58.7%	363	52.2%	72	53.7%
	Very Important	183	15.4%	132	19.0%	33	24.6%
The need to plug the vehicle in to recharge the battery	Not Important	106	8.9%	51	7.3%	8	5.9%
	Important	745	62.4%	424	60.8%	74	54.8%
	Very Important	343	28.7%	222	31.9%	53	39.3%
Concerns about limited access to plug-in locations	Not Important	191	16.1%	97	14.0%	16	11.9%
	Important	599	50.3%	318	45.8%	54	40.0%
	Very Important	400	33.6%	280	40.3%	65	48.1%
Availability of desirable vehicle size or style	Not Important	134	11.2%	81	11.7%	10	7.4%
	Important	580	48.6%	343	49.4%	72	53.3%
	Very Important	479	40.2%	271	39.0%	53	39.3%
Reliability of the vehicle	Not Important	173	14.5%	107	15.4%	14	10.4%
	Important	533	44.7%	318	45.7%	57	42.5%
	Very Important	486	40.8%	271	38.9%	63	47.0%
On-going maintenance and operative costs (including battery replacement)	Not Important	80	6.7%	34	4.9%	6	4.4%
	Important	566	47.5%	326	46.9%	59	43.7%
	Very Important	546	45.8%	335	48.2%	70	51.9%
The ability to carry heavy loads	Not Important	205	17.2%	83	11.9%	15	11.1%
	Important	502	42.1%	309	44.4%	58	43.0%
	Very Important	486	40.7%	304	43.7%	62	45.9%
Cells highlighted represent the highest percentage across income groups							

3.6.10 Impact on environment

Table 73 explains that respondents from all the income groups agreed with the statement that “cars, minivans, vans, pickups and SUVs are not an important source of air pollution anymore”, they also agreed with the statement that “cars, minivans, vans, pickups and SUVs are an important source of the greenhouse gases that many scientists believe are warming the earth’s climate”. Respondents from all the income groups disagreed with the statement that government rules require minivans, vans, pickups and SUVs to meet the same miles-per-gallon standards as passenger cars.

For the statement "government rules allow minivans, vans, pickups and SUVs to pollute more than passenger cars, for every gallon of gas used", there is a variation in the perceptions among all income groups. The income group of INR 25,000-50,000 and respondents who had monthly income above INR 1 lakh agreed with the above statement, however the respondents from monthly income group of INR 50,000-100,000 disagreed with the statement.

Respondents from all income groups agreed with the statement that exhaust from cars, minivans, vans, pickups and SUVs is an important source of pollution that causes asthma and makes asthma attacks worse.

Table 73: Breakup of impact on environment of various transport modes by income group

		Monthly income (INR)					
		25,000 – 50,000		50,000 – 100,000		Above 1 lakh	
		Count	Column N %	Count	Colum n N %	Count	Colum n N %
Cars, minivans, vans, pickups and SUVs are not an important source of air pollution anymore.	Agree	824	69.1%	448	64.2%	66	48.5%
	Disagree	307	25.7%	196	28.1%	57	41.9%
	Unsure	62	5.2%	54	7.7%	13	9.6%
Government rules allow minivans, vans, pickups and SUVs to pollute more than passenger cars, for every gallon of gas used.	Agree	529	44.5%	258	37.0%	50	37.0%
	Disagree	511	43.0%	313	44.8%	50	37.0%
	Unsure	149	12.5%	127	18.2%	35	25.9%
Cars, minivans, vans, pickups and SUVs are an important source of the greenhouse gases that many scientists believe are warming the earth's climate	Agree	620	52.0%	290	41.6%	75	55.1%
	Disagree	368	30.9%	223	32.0%	37	27.2%
	Unsure	204	17.1%	184	26.4%	24	17.6%
Government rules require minivans, vans, pickups and SUVs to meet the same miles-per-gallon standards as passenger cars.	Agree	398	33.4%	215	30.9%	41	30.1%
	Disagree	534	44.8%	274	39.4%	49	36.0%
	Unsure	259	21.7%	207	29.7%	46	33.8%
Exhaust from cars, minivans, vans, pickups and SUVs is an important source of the pollution that causes asthma and makes asthma attacks worse	Agree	532	44.5%	254	36.4%	72	52.9%
	Disagree	374	31.3%	255	36.5%	39	28.7%
	Unsure	289	24.2%	189	27.1%	25	18.4%
Cells highlighted represent the highest percentage across income groups							

3.6.11 Premium for electric vehicles

All the respondents from all the income groups were willing to pay a premium ranging from 10% to 20%. The majority of the respondents from all income groups were willing to pay 15% premium for electric vehicles (Table 74).

Table 74: Willing to pay a premium for electric vehicle

		Monthly income					
		25,000 – 50,000		50,000 – 100,000		Above 1 lakh	
		Count	Column N %	Count	Column N %	Count	Column N %
If very/somewhat interested in electric vehicles would you be willing to pay a premium to purchase an electric vehicle?	10%	343	29.5%	174	25.2%	39	29.1%
	15%	480	41.3%	323	46.8%	60	44.8%
	20%	339	29.2%	193	28.0%	35	26.1%
Cells highlighted represent the highest percentage across income groups							

3.7 Environmental preferences

In our routine life, transportation plays a crucial role. In today's world, people are travelling more for work and spending more time on vacations, and with growing prosperity, this trend is going to continue. Because of the increasing use of automobiles, the world is seeing pressure on fossil fuel supplies, fluctuating oil prices and adverse impact on the environment. This has increased the needfor electric vehicles and clean energy sources. Considering the scope of the study, there were were several objectives of this study but one of the prime objectives was to find an association between socio-demographic variables such as age group, gender, income status and education level with environment related factors.The results were analyzed and are summarized below.

3.7.1 Vehicular pollution and its impact on air quality

In Table 75,questions were asked related to the respondent's education level and vehicle emissions which affect the local air quality. It was observed that respondents who are currently pursuing their education are more sensitive towards vehicular pollution and understand that it is a major problem which is affecting the air quality. 53.3% respondents who are studying confirm that vehicle emissions affect local air and quality is becoming major problem; 43.1% of these respondents also confirm that it is a problem. However, only

3.6% of the respondents who are students shared that vehicle emissions do not cause degradation of local air quality.

32.2% of respondents who have completed their graduation degrees confirm that vehicle emissions which affect local air quality are becoming a major problem; 44.2% of these respondents also confirm that it is a problem. However, only 23.6% of the respondents shared that vehicle emissions do not cause degradation of local air quality.

32.7% respondents who have completed their master's degrees confirm that vehicle emissions affect local air, and quality is becoming a major problem; 49.2% such respondents also confirm that it is a problem. However, only 18.1% of these respondents shared that vehicle emissions are not a problem for local air quality.

Table 75: Vehicle emission by education level of the respondents

			"Vehicle emissions that affect local air quality"			Total
			Not a Problem	Problem	Major Problem	
Education	Bachelor's	Count	247	462	336	1045
		% within Education	23.6%	44.2%	32.2%	100.0%
	Master's	Count	159	432	287	878
		% within Education	18.1%	49.2%	32.7%	100.0%
	Student	Count	11	131	162	304
		% within Education	3.6%	43.1%	53.3%	100.0%
Total		Count	417	1025	785	2227
		% within Education	18.7%	46.0%	35.2%	100.0%
Cells highlighted in green represent the highest percentage across education levels						

3.7.2 Climate change

Table 76 highlights that 37% of respondents who have completed their graduation feel that vehicle emissions contribute to climate change and are becoming a major problem nowadays, for 53.4% of these respondents it is a problem, however, 9.6% of respondents who are graduates feel that vehicle emissions are not a contributing factor for climate change.

36.3% of respondents who have completed their master's degrees feel that vehicle emissions contribute to climate change and are becoming a major problem nowadays, for 52.3% such respondents it is a problem, however, 11.5% of respondents who are graduates feel that vehicle emissions are not a contributing factor for climate change.

52.3% of respondents who have completed their master's degree feel that vehicle emissions contribute to climate change and are becoming a major problem nowadays, for 41.8% such respondents it is a problem, however, 5.9% of respondents who are master's degree level feel that vehicle emissions are not a contributing factor to climate change. It can be concluded from the data that students are more sensitive towards the environment and only a small number (5.9%) of such respondents feel that vehicle emissions are not the reason for climate change.

Table 76: Perceptions of vehicle emission by education level of the respondents

			Vehicle emissions that contribute to climate change			Total
			Not a Problem	Problem	Major Problem	
Education	Bachelor's	Count	100	558	386	1044
		% within Education	9.6%	53.4%	37.0%	100.0%
	Master's	Count	101	460	319	880
		% within Education	11.5%	52.3%	36.3%	100.0%
	Student	Count	18	127	159	304
		% within Education	5.9%	41.8%	52.3%	100.0%
Total		Count	219	1145	864	2228
		% within Education	9.8%	51.4%	38.8%	100.0%

Cells highlighted represent the highest percentage across education levels

3.7.3 Vehicular pollution – effect on health

Table 77 highlights that 49.1% of respondents who are graduates agreed with the statement that "exhaust from cars, minivans, vans, pickups and SUVs is an important source of the pollution that causes asthma and makes asthma attacks worse," and 30.8% of respondents disagreed with this statement. However, 22% of such respondents are unsure if "vehicular pollution causes asthma and makes asthma attacks worse".

36% of respondents who have completed their master's degree agreed with the statement that "exhaust from cars, minivans, vans, pickups and SUVs are an important source of the pollution that causes asthma and makes asthma attacks worse," and 36.5% of these respondents disagreed with this statement. However, 27.5% of such respondents are unsure if "vehicular pollution causes asthma and makes asthma attacks worse".

62.3% respondents who are currently studying agreed with the statement that "exhaust from cars, minivans, vans, pickups and SUVs are an important source of the pollution that causes asthma and makes asthma attacks worse," and 20.5% of such respondents disagreed with this statement. However, 17.1% of such respondents are unsure if "vehicular pollution causes asthma and makes asthma attacks worse".

Table 77: Effect on human health caused by vehicular pollution

			"Exhaust from cars, minivans, vans, pickups and SUVs is an important source of the pollution that causes asthma and makes asthma attacks worse"			Total
			Agree	Disagree	Unsure	
Education	Bachelor's	Count	509	300	228	1037
		% within Education	49.1%	28.9%	22.0%	100.0%
	Master's	Count	315	320	241	876
		% within Education	36.0%	36.5%	27.5%	100.0%
	Student	Count	182	60	50	292
		% within Education	62.3%	20.5%	17.1%	100.0%
Total		Count	1006	680	519	2205
		% within Education	45.6%	30.8%	23.5%	100.0%
Cells highlighted represent the highest percentage across education levels						

3.7.4 Environmental pollution – gender perception on government rules

Table 78 explains that 40.7% of male respondents agree with the statement that "government rules allow minivans, vans, pickups and SUVs to pollute more than passenger cars, for every gallon of gas used", while 44.4% disagree with it. However, 14.8% of male respondents were unsure whether "government rules allow minivans, vans, pickups and SUVs to pollute more than passenger cars, for every gallon of gas used" or not.

44.4% of female respondents agree with the statement that "government rules allow minivans, vans, pickups and SUVs pollute more than passenger cars, for every gallon of gas used", while 38.0% disagree with it. However, 17.7% of female respondents were unsure whether "government rules allow minivans, vans, pickups and SUVs to pollute more than passenger cars, for every gallon of gas used" or not.

Table 78: Gender perception related to government rule

			"Government rules allow minivans, vans, pickups and SUVs to pollute more than passenger cars, for every gallon of gas used".			
			Agree	Disagree	Unsure	Total
Gender	Male	Count	502	547	184	1233
		% within gender	40.7%	44.4%	14.9%	100.0%
	Female	Count	424	363	169	956
		% within gender	44.4%	38.0%	17.7%	100.0%
Total		Count	926	910	353	2189
		% within gender	42.3%	41.6%	16.1%	100.0%
Cells highlighted represents the highest percentage across gender						

3.7.5 Environment-friendly vehicle – willingness to pay premium

It is observed from various sources that people have started thinking about the environment and climate change. They want to contribute in some way. Table 79 explains the relationship between income level and willingness to pay a premium for environmentally friendly vehicles. Respondents were asked if they would pay a premium for electric vehicle. 29.5% of the respondents who had a monthly income between INR 25,000-50,000 are willing to pay a premium of 10%, 41.3% of such respondents can pay a15% premium and 29.2% of such respondents can pay up to 20% premium.

25.2% of the respondents who had a monthly income between INR 50,000-100,000 are willing to pay a premium of 10%, 46.8% of such respondents can pay a15% premium and 28% of such respondents can pay up to 20% premium.

29.1% of the respondents who have a monthly income of INR 1 lakh are willing to pay a premium of 10%, 44.8% of such respondents can pay a15% premium and 26.1% of such respondents can pay up to 20% premium.

Table 79: Willingness to pay a premium for environment friendly vehicle

			Willingness to pay a premium for environment friendly vehicle			Total
			10%	**15%**	**20%**	
Monthly income	25,000 – 50,000	Count	343	480	339	1162
		% within income level	29.5%	41.3%	29.2%	100.0%
	50,000 – 100,000	Count	174	323	193	690
		% within income level	25.2%	46.8%	28.0%	100.0%
	Above 1 lakh	Count	39	60	35	134
		% within income level	29.1%	44.8%	26.1%	100.0%
Total		Count	556	863	567	1986
		% within Education	28.0%	43.5%	28.5%	100.0%
Cells highlighted represents the highest percentage across income levels						

Chapter 4

Consumer Survey Outcomes and Recommendations

4.1 Findings and conclusion

2281 respondents were surveyed for their views regarding EVs. The key findings and recommendations are mentioned in this section.

1. It was found that young people can be targeted for this EV technology. Age group of 18-30 will be the perfect age group to target, because of their acceptance level and purchasing power. We found that 55.5% who fall under the age group of 18-30 can become potential customers in the near future.

2. It was observed that 92.8% of people who responded to our questionnaire do not own an electric vehicle.

3. People were interested in buying electric vehicles, but they were waiting for its availability in the market. More than 69% of such respondents can be targeted under this segment.

4. There are around 85.1% respondents who do not travel more than 60 km per day. If people are made aware of the EV technology and its driving range, this group can be converted into a prospective buyer's group of electric vehicles.

5. Most of the respondents shared that they get to know about new technologies from various sources like – television, newspapers, internet and magazines. Very few respondents say that they get technological updates from radio and other sources.

6. More than 84% of respondents shared that they purchase vehicles after discussing it with their family, or take a joint decision.

7. More than 80% of respondents shared that they would like to wait for the new technology to receive favorable customer feedback before purchasing

a new vehicle. They prefer to be followers and take a decision about purchasing after the product establishes a good reputation in the market.

8. Most of the respondents shared that fuel efficiency (83%), safety (63%), vehicle power (62%), purchase price (61%) and reliability (49%) are very important, however some of them also shared that vehicle size (46.4%), operating cost (53.3%), reputation (54.9%), fuel type (61.9%), vehicle emission and pollution (61.2%) play an important part in deciding whether to purchase the vehicle.

9. Approximately 50% shared that traffic congestion is a problem for them. Respondents are also facing problems of traffic noise (63.6%) and vehicle emission (51%) at several locations which affect their quality of life. Respondents also revealed that speeding traffic (48.7%) and heavy import of crude oil (47.5%) have also become a problem for them directly or indirectly.

10. It was analyzed that people are ready to buy an electric vehicle if it can reduce their operating costs. They also shared that it is eco-friendly and reduces the country's dependence on import of fossil fuel.

11. Approximately 61% of respondents shared they do not agree with the statement that "cars, minivans, vans, pickups and SUVs are an important source of air pollution" and they also agree with the statement that "cars, minivans, vans, pickups and SUVs are an important source of the greenhouse gases that many scientists believe are warming the earth's climate". Around 45% of the respondents shared that "exhaust from cars, minivans, vans, pickups and SUVs is an important source of pollution that causes asthma and makes asthma attacks worse".

12. Through this survey, it was found that 95% of people are ready to pay a premium for electric vehicles.

13. Based on the data, East is a suggested test or launch market for new technologies or vehicles.

Findings from Cross Tabulation Observations

Analysis of responses by region

1. In South India, a majority of people travel upto 30 km in a day. In East India, people travel shorter distances to complete their routine work, almost 75% of the respondents shared that they travel upto 20 km in a day to complete their work.

2. In West India and North India, the situation is different from the other two regions. In these two regions, people travel more to complete their work. Data revealed that approximately 42% of respondents in the West region and 37% of respondents in the North region travel more than 30 km in a day for their routine work. Male respondents were travelling more in all the regions compared to women.

3. In the South region, the internet is the main source of information followed by television and newspapers. Similarly, in the North region, respondents shared that the major information source is the internet followed by newspapers, television, magazines and radio.

4. In the West region, television and internet are the main information sources. The region has other important information sources as well such as newspapers, magazines and radio. In East India, the major source of information is television followed by newspapers, the internet, magazines, word of mouth and radio.

5. In the South region, people take joint decisions while purchasing any new vehicle, however in the other three regions, they consult with elders while taking the decision about purchasing any vehicle. It can be concluded from the above data that across regions, every member of the family has a say in the decision while purchasing a vehicle for the family.

6. Respondents from the East region shared that they were among the first to purchase a new vehicle. Respondents from the South and the West region shared that they would wait to read the reviews and then buy it if the reviews are favorable. However the respondents from the North region shared that they will wait until the new technology is widely accepted and proven before purchasing the new vehicle. Based on the data, East is a suggested test or launch market for new technologies or vehicles.

7. Respondents from the East region are very interested in purchasing an electric motor vehicle once they become easily available in the next couple of years while respondents from the other three regions were skeptical about purchasing such electric vehicles. They were "somewhat interested" in purchasing such vehicles.
8. Respondents from the East region shared that the State and Central Government should encourage electric vehicles only for public transport, however respondents from the other three regions shared that the State and Central Government should encourage electric vehicles for both public and private transport by focusing on transport policy.
9. Approximately 15.4% respondents from the North region and 14.8% respondents from the West region did not find the current policies sufficient to promote electric vehicles in the market.
10. Factors such as fuel efficiency, reliability and vehicle power were very important for purchasing of a personal vehicle in all regions. For South, North and West region safety is a very important factor, however for East region respondents, it is an important factor. Vehicle size along with the market reputation of a particular vehicle is an important factor which influences the purchasing habits of respondents.
11. For a better understanding of the issues related to the transport system, questions focusing on the problems faced by the respondents were asked. Various variables were included such as "congestion and noise; vehicle emissions and pollution; traffic speed; global warming; fuel import" etc. These variables highlighted the preferences of respondents while buying any vehicle.
12. "Saving money on the cost of operation (using electricity rather than gasoline)" is not important for the North and East region, however, it is important for the South and West regions. Respondents from all regions were concerned about the environmental impact of vehicles and highlighted that electric vehicle will reduce pollution in the near future.
13. Respondents from West, North and East India felt that "it is important to reduce the impact of transport on the environment", however, respondents

from South India felt that “it is very important to reduce the impact of transport on the environment”.

14. Respondents from South India and East India feel that “it is important to reduce dependence on gasoline”. Respondents from North India and East India feel that “it is important to drive a vehicle with more advanced or innovative technology”.
15. The common issues that came out were “recharge stations, higher price, availability of desirable vehicle size, reliability, on-going maintenance, operating costs and the ability to carry occasionally heavy loads” – these are the issues that are restricting the demand for EVs in the Indian market.
16. For East India respondents, it is not important to purchase EVs by paying more than for a comparable conventional vehicle, for the rest of the regions, i.e. South, West and North it is important to pay more for an EV than for a comparable conventional vehicle.
17. For South India respondents, reliability of the vehicle is important, however, for Northern and Eastern regions, vehicle size is important. Respondents from West India were more concerned about limited access to plug-in locations and considered this an important factor.
18. Maintenance and the operating cost were very important factors for the West region and an important factor for North and East region. However, none of the regions were concerned about the heavy load capacity of EVs.
19. More than 50% of the respondents from West India, North India and East India agree that “cars, minivans, vans, pickups and SUVs are not an important source of air pollution anymore” while South India respondents disagree with the statement.
20. 51.8% respondents from West India agree that “government rules allow minivans, vans, pickups and SUVs to pollute more than passenger cars, for every gallon of gas used” while 55.6% disagree with the statement from East India. Majority of the respondents from South and North India neither agreed nor disagreed with that statement.
21. 65.8% respondents from South India and 72.8% respondents from East India agreed that “cars, minivans, vans, pickups and SUVs are an important source of the greenhouse gases that many scientists believe are

warming the earth's climate" while majority of the respondents from West India and North India neither agreed nor disagreed with that statement.

22. Respondents from all regions neither agreed nor disagreed with the statement that "government rules require minivans, vans, pickups and SUVs to meet the same miles-per-gallon standards as passenger cars" in the majority. This clearly shows that they were not worried about government policies.

23. 66.1% respondents from South India and 72.2% respondents from East India agreed that "exhaust from cars, minivans, vans, pickups and SUVs are an important source of pollution that causes asthma and makes asthma attacks worse" while majority of the respondents from West India and North India neither agreed nor disagreed with that statement.

24. All respondents from all regions were willing to pay a premium ranging from 10% to 20%. The majority of the respondents from the South region and North region were willing to pay 15% premium for electric vehicles. Respondents from the East region were willing to pay 10% premium and West region respondents were willing to pay 20% premium.

Analysis of responses by cities

1. The highest number of respondents who travel upto 10 km in a day, were from Hyderabad. Manesar respondents travel up to 20 km in a day. This was because most of the respondents were living in Gurgaon and prefer to commute on a daily basis.

2. Bengaluru had the highest number of respondents who travel up to 30 km in a day. Respondents from Pune shared that they travel up to 60 km in a day. Respondents from Mumbai travel more than 60 km in a day to complete their routine work.

3. It can be inferred from that data that respondents from all the cities can be targeted for the electric vehicle, because the travel distance within the cities can be easily covered by electric vehicle.

4. In Guwahati, elders take the decisions for most of the respondents, in Faridabad the decision is influenced by children and in Kochi, a joint decision is taken for vehicle purchases.

5. Respondents from Guwahati are ready to purchase vehicles once they are made available in the market. Respondents from Pune will wait to read reviews of it and then buy it if the reviews are favorable.

6. Respondents from Manesar and Faridabad prefer to wait until the new technology is widely accepted and proven in the market before considering it.

7. Respondents from all the cities shared that they are ready to purchase an electric vehicle once they become easily available in the market in the next couple of years.

8. Respondents from Guwahati and Gurgaon were "interested" and "somewhat interested" in purchasing an electric vehicle in the near future. Respondents from small cities were not very interested in purchasing electric vehicles.

9. Respondents from all the cities were in favor of encouraging the State and Central Government to promote electric vehicles in the market. Respondents from Chennai and Lucknow were strongly in favor of encouraging transport policies for both public and private transport.

10. Respondents from Guwahati were in favor of promoting policies only for public transport. Respondents from Faridabad shared that the existing policies are not worth promoting.

11. Respondents shared that traffic congestion that they experience while driving is a major problem for Thiruvananthapuram, a problem for Mumbai and for Guwahati it is not a problem.

12. Respondents from Chennai shared that traffic noise that they hear at home, work or school is a major problem for them, for Guwahati respondents it was a problem, and for Bengaluru respondents it was not a problem.

13. Respondents from Hyderabad shared that vehicle emissions that affect local air quality area major problem for them, for Jaipur respondents it was a problem, for Guwahati it was not a problem.

14. Data shows that for Bengaluru respondents, vehicle emissions that contribute to climate change were not a problem, for Guwahati

respondents it was a problem, for Hyderabad respondents it was a major problem.

15. Respondents shared that they feel unsafe because of speeding traffic. For Kochi respondents it was a major problem, for Jaipur respondents it was a problem, however Gurgaon respondents shared that it is not a problem.

16. Respondents from Pune shared that fuel import from foreign countries is not a problem for them, for Guwahati respondents it is a problem and for Bengaluru respondents it is a major problem.

17. Saving money on the cost of operation is not important for the respondents from Guwahati, for Jaipur respondents it is an important factor and for Hyderabad respondents it is a very important factor. Reduced impact on the environment is a very important factor for the respondents from Kochi, it is an important factor for the respondents from Gurgaon. However, it is not an important factor for Bengaluru respondents.

18. Reduced dependence on gasoline is a very important factor for Mumbai and Thiruvananthapuram respondents, for Jaipur respondents it is an important factor, however, for Guwahati respondents it is not an important factor. Driving a vehicle with more advanced or innovative technology is very important for Bengaluru respondents, for Gurgaon respondents it is an important factor and for Kochi it is not an important factor.

19. The common issues were "recharge stations, higher price, availability of desirable vehicle size, reliability, on-going maintenance, operating costs and the ability to carry occasionally heavy loads" which are restricting the demand for EVs in the Indian market.

20. Respondents from most of the cities agreed with the statement that cars, minivans, vans, pickups and SUVs are not an important source of air pollution anymore. Respondents from Kochi disagreed with the statement and shared that that cars minivans, vans, pickups and SUVs are an important source of air pollution.

21. Respondents from Coimbatore agreed with the statement that government rules allow minivans, vans, pickups and SUVs to pollute more than passenger cars, for every gallon of gas used, however respondents from Faridabad disagreed with this statement.

22. Respondents from Hyderabad agreed with the statement that cars, minivans, vans, pickups and SUVs are an important source of the greenhouse gases that many scientists believe are warming the earth's climate, however respondents from Faridabad disagreed with the statement.

23. Respondents from Chennai agreed with the statement that exhaust from cars, minivans, vans, pickups and SUVs is an important source of pollution that causes asthma and makes asthma attacks worse, however respondents from Faridabad disagreed with the above-mentioned statement. Bengaluru respondents were unsure about the above-mentioned statement.

24. Respondents from all the cities were willing to pay a premium for electric vehicles. It was revealed that respondents from Guwahati were ready to pay 10% premium over the conventional vehicle to save the environment, Gurgaon respondents were ready to pay 15% premium and Mumbai respondents were ready to pay 20% premium.

Analysis of responses by age groups

1. Data revealed that the age group of 18-30 and 31-60 travel upto 20 km per day for their routine work, however, the age group of above 61 years travel upto 60 km or more in a day.

2. Considering the data on per day drive it is clear that the group of 18-30 years if motivated properly and given sufficient services, can think of using electric vehicles for their short travels.

3. Television is the important source of information among the age group of 18-30 and above 61, however, among the age group of 31-60, the newspaper is the main source of information followed by the internet.

4. Data revealed that all the age groups, i.e. 18-30, 31-60, 61 and above discuss the decision with family members while taking a decision about purchasing any vehicle.

5. Respondents from all the age groups shared that they would prefer to wait to read the reviews of a new vehicle and then will buy it if the reviews are favorable.

6. Respondents from all the age groups are interested in buying an electric vehicle once they become easily available in the next couple of years. The age group 18-30 has higher percentage of respondents compared to the other age groups who are very interested in buying such vehicles.

7. Respondents from all the age groups felt that the State and Central Government should encourage electric vehicles for both public and private transport though amendments in transport policy. Approximately 25% of the respondents who were above 61 years felt that the policies defined by State and Central Government are not sufficient for both the sectors.

8. Factors such as fuel efficiency, vehicle power, reliability and fuel type are very important for all the age groups. Safety is a very important factor for the age group of 18-30, but an important factor for the other two age groups. Purchase price, reputation of the vehicle, vehicle emissions and pollution are important factors for all the age groups.

9. Data highlights that saving money on the cost of operation (using electricity rather than gasoline) is not an important factor for the age group of 31-60 and above 61 years old people, however, it is an important factor for the age group of 18-30.

10. Respondents from all the age groups were concerned about the environmental impact of transport, and highlighted that electric vehicle will reduce the pollution in the coming future.

11. For the age group of 18-30 and 31-60, it is important to reduce the dependence on gasoline, however, respondents from above 61 years do not consider it an important factor. Respondents from all the age groups shared that "it is important to drive a vehicle with more advanced or innovative technology".

12. Various variables such as purchase price, recharge options, recharging stations, availability of vehicle size and vehicle reliability are important factors for respondents from all the age groups. On-going maintenance and operative costs and load capacity are very important factors for the age group of 18-30, however these factors are important to the rest of the age groups.

13. Respondents from all the age groups agreed with the statement that "cars, minivans, vans, pickups and SUVs are not an important source of air pollution anymore".

14. Respondents from the age group of 18-30 and above 61 years agreed with the statement that government rules allow minivans, vans, pickups and SUVs to pollute more than passenger cars, for every gallon of gas used, however respondents from the age group of 31-60 disagree with the statement.

15. Respondents from all the age groups agreed with the statement that cars, minivans, vans, pickups and SUVs are an important source of the greenhouse gases that many scientists believe are warming the earth's climate.

16. Respondents from all the age groups disagreed with the statement that government rules require minivans, vans, pickups and SUVs to meet the same miles-per-gallon standards as passenger cars.

17. Respondents from the age group of 18-30 and above 61 years agreed with the statement that exhaust from cars, minivans, vans, pickups and SUVs is an important source of pollution that causes asthma and makes asthma attacks worse, however age group of 31-60 disagreed with the above-mentioned statement.

18. Respondents from all the age groups were willing to pay a premium for electric vehicles. It was revealed that respondents were ready to pay 15% premium over the conventional vehicle to save the environment.

Analysis of responses by gender

1. There is no significant difference between travel distance covered by male and female respondents. It is evident from the table that approximately 71% male respondents and 63% female respondents travel upto 30 km in a day for their routine work.

2. Data highlighted that there is no significant difference in the information sources for both the genders. The main source of information for male and female respondents is the internet.

3. Television followed by newspapers and magazines were the other important information sources for male respondents. On the other hand, newspapers followed by television and magazines were the other important information sources for female respondents.

4. Elders influence the purchasing decision for the both the genders. Women respondents give higher priority to children and take joint decisions compared to male respondents.

5. Both male and female respondents wanted to wait for the reviews of the vehicle and then buy it if the reviews are favorable.

6. Respondents were ready to purchase an electric vehicle once they become easily available in the next couple of years. Female respondents were more likely to buy such vehicles compared to male respondents.

7. Both male and female respondents shared that the State and Central Government should encourage transport policy for electric vehicles.

8. Female respondents were more concerned about traffic congestions than male respondents. They expressed that traffic congestions are problematic for them.

9. Traffic noise was a problem for both male and female respondents. Male respondents expressed that vehicle emissions, which affect local air quality are a problem for them.

10. Data highlighted that both male and female respondents agree with the statement that "cars, minivans, vans, pickups and SUVs are not an important source of air pollution anymore" because of the changed technology.

11. However, male respondents agreed with the statement that "cars, minivans, vans, pickups and SUVs are an important source of the greenhouse gases that many scientists believe are warming the earth's climate". For the rest of the variables, no clear majority was observed in the analysis.

12. For both male and female respondents "reduced impact on the environment" is an important variable for them and it is important for them to drive a vehicle with more advanced or innovative technology.

13. Data highlights that "higher purchase price for EV than for a comparable conventional vehicle" is an important factor for both male and female respondents and also revealed that the need to plug the vehicle in to recharge the battery is an important factor for them. Female respondents were concerned about the vehicle size and considered it an important factor. For the rest of the variables, no clear majority was observed in the analysis.

14. Fuel efficiency, safety, vehicle power and reliability are very important factors for both male and female respondents. Purchase price, vehicle size, vehicle reputation, vehicle emission and pollution are important factors for both male and female respondents. Males were concerned about fuel type and consider this very important.

15. Both male and female respondents were willing to pay a premium for an electric vehicle. It was revealed that respondents were ready to pay 15% premium over the conventional vehicle to save the environment.

Analysis of responses by education level

1. Data revealed that the percentage of students who travel less than 10 km in a day is higher compared to those respondents who had completed their bachelor's and master's degree.

2. The respondents who had completed their bachelor's and master's degree were travelling upto 20 km in the day to reach their offices. From the data, it can be concluded that most of the respondents from all the age groups were travelling upto 60 km in a day.

3. Respondents with different education backgrounds had different sources of information. Television was the main source of information for those respondents who had completed their bachelor's degree. For respondents who had completed their master's degree, radio was the main source of information. However, students were mostly dependent on word of mouth and other information sources for knowing about the latest developments around them.

4. Data revealed that elders influence the purchasing decisions at all the education levels. For themajority of the respondents, elders take the

decision for any vehicle purchase. However students shared that the decision is taken jointly.

5. Respondents from all the age groups wanted to wait for reading the reviews and then buy the vehicle only if the reviews are favorable. Students wanted to wait until the new technology is widely accepted and proven before considering it.

6. Respondents were ready to purchase an electric vehicle once they become easily available in the next couple of years. Majority of respondents from all the education groups were "somewhat interested" in purchasing an electric motor vehicle once they become easily available in the next couple of years.

7. Respondents from all the education groups felt that the State and Central Government should encourage electric vehicles for both public and private transport though amendments in transport policy. Only 9% of the respondents who were students shared that the policies defined by State and Central Government are not sufficient for both the sectors.

8. For all the education groups, fuel efficiency, safety, vehicle power and reliability were very important factors. Vehicle purchase price and vehicle size were important factors for all the respondents from all the educationgroups.

9. Operating cost of thevehicle was very important to those respondents who had completed their bachelor's and master's degrees, however for students it was just an important factor.

10. Students were more concerned about vehicle emission and pollution and considered this a very important factor while other groups of respondents considered it an important factor.

11. Respondents from all the education groups shared that traffic congestion while driving, speeding traffic and traffic noise at home, work or school are a problem for them and this has created a lot of issues in society.

12. The other factors such as vehicle emissions that affect local air quality and importing fuel from foreign countries were a major problem for the students, however for the other education groups, it was just a problem.

13. Data highlighted that saving money on the cost of operations was an important factor for those respondents who had completed their bachelor's degree or are studying, but it was not an important factor for those respondents who had completed their master's degrees.

14. Respondents from all the education groups shared that it is important to drive a vehicle with more advanced or innovative technology and reduce the impact on the environment by using electric vehicles which will also reduce dependence on gasoline.

15. Variables such as "higher purchase price than for a comparable conventional vehicle, the need to plug the vehicle in to recharge the battery, concerns about limited access to plug-in locations, availability of desirable vehicle size or style, on-going maintenance & operative costs and the ability to carry heavy loads" are important factors for all respondents from all age groups.

16. Student respondents were concerned about vehicle reliability and shared that it is a very important factor for them, however the other education groups shared that it is just an important factor.

17. Respondents who have completed their bachelor's and master's degrees agreed with the statement that cars, minivans, vans, pickups and SUVs are not an important source of air pollution anymore, but the respondents who were students disagreed with this statement.

18. Respondents from all the education levels agreed with the statement that cars, minivans, vans, pickups and SUVs are an important source of the greenhouse gases that many scientists believe are warming the earth's climate.

19. All the respondents from all the education groups were willing to pay a premium ranging from 10% to 20%. The majority of the respondents from all the education groups were willing to pay 15% premium for electric vehicles.

Analysis of responses by income group

1. Respondents with a higher income can afford to travel more to finish their work. Approximately 82% respondents who fall under the monthly income bracket of INR 25,000-50,000 travel upto 30 km.

2. Similarly 59% of the respondents who fall under the monthly income bracket of INR 50,000-100,000 travel upto 30 km and 51% of the respondents who have a monthly income of more than INR 1 lakh per month undertake travel upto 30 km in a day.

3. Data revealed in Table 64 that there is a difference between sources of information among income groups. In monthly income group of INR 25,000-50,000 television is the main source of information followed by newspapers and internet.

4. For the monthly income groups of INR 50,000-100,000 and above INR 1 lakh, the internet remains the prime source of information, followed by newspapers, television, magazines and radio.

5. Respondents from monthly income group of INR 25,000-50,000 and INR 50,000-100,000 consult with elders while taking the decision about purchasing any vehicle. The respondents who had an income of more than INR 1 lakh per month were taking a joint decision about purchasing any vehicle.

6. In all the income groups, respondents shared that they will wait to read the reviews and then will buy it if the reviews are favorable.

7. Respondents also wanted to wait until the new technology is widely accepted and proven in the market. Data also revealed that people from the income group of INR 25,000-50,000 are ready to purchase a new vehicle compared to respondents from other income groups.

8. Respondents from all income groups were very interested in purchasing an electric motor vehicle once they become easily available in the next couple of years but they were "somewhat interested" in purchasing such vehicles.

9. There is no significant difference which exists between the respondents who are not planning to purchase the electric vehicle in all the income groups.

10. Respondents from all the income groups shared that the State and Central Government should encourage electric vehicles for public transport. Respondents also shared that the State and Central Government should encourage electric vehicles for private transport.

11. Factors such as fuel efficiency, safety, vehicle power and reliability of the vehicle were very important for purchasing of a personal vehicle in all the income groups. Vehicle size, emissions and pollution are important factors for all the income groups.

12. There is a difference between the perceptions of respondents with regard to fuel type and operating cost of the vehicles. For the monthly income group of INR 25,000-50,000 and above INR 1 lakh, they are very important factors, however for the monthly income group of INR 50,000-100,000 it is just an important factor.

13. To understand the issues related to the transport system, questions focusing on the problems faced by the respondents were asked. Various variables were included such as, "congestion and noise; vehicle emissions and pollution; traffic speed; global warming; fuel import" etc. These variables highlighted the preferences of respondents while buying any vehicle. The majority of the respondents expressed that these factors are a problem for them.

14. Data highlights that for the following statement: "saving money on the cost of operation (using electricity rather than gasoline)", respondents from monthly income group of INR 25,000-50,000 were not able to decide whether it is an important factor or not.

15. Respondents from the monthly income group of INR 50,000-100,000 shared that this is not an important factor for them, however the respondents who had monthly income more than INR 1 lakh considered it an important factor to consider while purchasing any vehicle.

16. Respondents from all the income groups shared that it is important to drive a vehicle with more advanced or innovative technology and reduce the impact on the environment by using electric vehicles which will also reduce dependence on gasoline.

17. Respondents from the entire income group shared that the variables such as higher purchase price than for a comparable conventional vehicle, recharging facilities, and availability of desirable vehicle size or style are important factors for them.

18. Respondents from the monthly income group of INR 25,000-50,000 and INR 50,000-100,000 were concerned about there liability of the vehicle, on-going maintenance and operative costs, and ability to carry heavy loads – these were important variables for purchasing the electric vehicle, but for the respondents who had an income of more than INR 1 Lakh these factors were very important.

19. Respondents from all the income groups agreed with the statement that "cars, minivans, vans, pickups and SUVs are not an important source of air pollution anymore", they also agreed with the statement that cars, minivans, vans, pickups and SUVs are an important source of the greenhouse gases that many scientists believe are warming the earth's climate.

20. Respondents from all the income groups disagreed with the statement that government rules require minivans, vans, pickups and SUVs to meet the same miles-per-gallon standards as passenger cars.

21. For the statement that government rules allow minivans, vans, pickups and SUVs to pollute more than passenger cars, for every gallon of gas used, there is a variation on the perceptions among all the income groups.

22. The monthly income group of INR 25,000-50,000 and respondents who hadmonthly income above INR 1 lakh agreed with the above statement, however the respondents from monthly income group of INR 50,000-100,000 disagreed with the statement.

23. Respondents from all the income groups agreed with the statement that exhaust from cars, minivans, vans, pickups and SUVs is an important source of the pollution that causes asthma and makes asthma attacks worse.

24. All the respondents from all the income groups were willing to pay a premium ranging from 10% to 20%. The majority of the respondents from

all the income groups were willing to pay a 15% premium for electric vehicle.

Analysis of responses by environment

1. Respondents were asked questions about vehicle emissions, which affect the local air quality and data was analyzed by education levels. It was observed that respondents who are currently pursuing their education are more sensitive towards vehicular pollution and understand that it is a major problem which is affecting the air quality.

2. 53.3% respondents who are studying confirm that vehicle emissions affect local air and quality is becoming a major problem; 43.1% such respondents also confirm that it is a problem. However, only 3.6% of the respondents who are students shared that vehicle emission is not a problem for local air quality.

3. 32.2% respondents who have completed their graduation degrees confirm that vehicle emissions affect local air and quality is becoming a major problem; 44.2% such respondents also confirm that it is a problem. However, only 23.6% of the respondents shared that vehicle emission is not a problem for local air quality.

4. 32.7% respondents who have completed their master's degrees confirm that vehicle emissions affect local air and quality is becoming a major problem; 49.2% such respondents also confirm that it is a problem. However, only 18.1% of such respondents shared that vehicle emission is not a problem for local air quality.

5. 37% respondents who have completed their graduation feel that vehicle emissions contribute to climate change and have become a major problem nowadays, for 53.4% such respondents it is a problem, however, 9.6% of respondents who are graduates feel that vehicle emission is not a contributing factor to climate change.

6. In the same line, 36.3% respondents who have completed their master's degrees feel that vehicle emissions contribute to climate change and have become a major problem nowadays, for 52.3% such respondents it is a

problem, however, 11.5% of respondents who are graduates feel that vehicle emission is not a contributing factor to climate change.

7. In the same line, 52.3% respondents who have completed their master's degrees feel that vehicle emissions contribute to climate change and have become a major problem nowadays, for 41.8% such respondents it is a problem, however, 5.9% of respondents feel that vehicle emission is not a contributing factor to climate change.

8. It can be concluded from the data that students are more sensitive towards the environment and a small number (only 5.9%) of such respondents feels that vehicle emission is not the reason for climate change.

9. 49.1% respondents who are graduates agreed with the statement that "exhaust from cars, minivans, vans, pickups and SUVs is an important source of the pollution that causes asthma and makes asthma attacks worse," and 30.8% disagreed with the statement. However, 22% of such respondents are unsure if "vehicular pollution causes asthma and makes asthma attacks worse".

10. 36% respondents who have completed their master's degrees agreed with the statement that "exhaust from cars, minivans, vans, pickups and SUVs are an important source of the pollution that causes asthma and makes asthma attacks worse," and 36.5% such respondents disagreed with the statement. However, 27.5% of such respondents are unsure if "vehicular pollution causes asthma and makes asthma attacks worse".

11. 62.3% respondents who are currently studying agreed with the statement that "exhaust from cars, minivans, vans, pickups and SUVs is an important source of the pollution that causes asthma and makes asthma attacks worse," and 20.5% such respondents disagreed with the statement. However, 17.1% of such respondents are unsure if "vehicular pollution causes asthma and makes asthma attacks worse".

12. 40.7% of male respondents agree with the statement that "government rules allow minivans, vans, pickups and SUVs to pollute more than passenger cars, for every gallon of gas used" while 44.4% disagree with it. However, 14.8% of male respondents were unsure whether "government rules allow minivans, vans, pickups and SUVs to pollute more than passenger cars, for every gallon of gas used" or not.

13. 44.4% of female respondents agree with the statement that "government rules allow minivans, vans, pickups and SUVs to pollute more than passenger cars, for every gallon of gas used" while 38% disagree with it. However, 17.7% of female respondents were unsure whether "government rules allow minivans, vans, pickups, and SUVs pollute more than passenger cars, for every gallon of gas used" or not.

14. It is observed from the data that people have started thinking about the environment and climate change. They want to contribute in some way. Respondents were asked if they can pay a premium for an EV. 29.5% of the respondents who have a monthly income between INR 25,000-50,000 are willing to pay premium of 10%, 41.3% of such respondents can pay a15% premium and 29.2% of such respondents can pay upto 20% premium.

15. 25.2% of the respondents who have a monthly income between INR 50,000-100,000 are willing to pay a premium of 10%, 46.8% of such respondents can pay a15% premium and 28% of such respondents can pay upto 20% premium.

16. 29.1% of the respondents who have a monthly income of INR 1 lakh are willing to pay premium of 10%, 44.8% of such respondents can pay a15% premium and 26.1% of such respondents can pay upto 20% premium.

The upcoming challenges

1. The market share of EV in the auto industry is small with respect to conventional vehicles. There are not many manufacturers in either the two wheelers or four wheelers segment. The Smart Cities project is expected to attract investment under its green transport category.

2. Convincing the consumers to pay a premium more than 15% for electric vehicle may become a hurdle and government assistance to towards this will be helpful in promoting the uptake of electric vehicles. Government support will also be required in the development of EV related infrastructure, like charging stations, workshops throughout the country, especially in the metro cities.

3. The Government of India can think of annual fuel savings of INR13,000 – 14,000 cr in 2020 by targeting 7 million sales of EVs.

4. The discussants added many more points. They touched upon ecological concerns, marketing strategies, policies, the resale value of batteries and EVs, and gender sensitivity, the concerns of Residents Welfare Associations, policy decisions, support from the Government, R& D, and all pertinent issues relating to EVs in India.

Mass use of electric vehicles has two major advantages- lower dependency on imported fuel and reduction in pollution level. Use of thermal power helps with the first point as we have a good reserve of lignite and coal; but it certainly does not have the benefits of electric cars regarding pollution level. On the other hand, hydropower does not come without permanent damage to ecology and increasing the risk of natural disasters. Though nuclear power is free from these disadvantages, concerns regarding safety and political opposition will remain major roadblocks. But, India has huge potential in wind and solar energy. We are already among top five or six countries and probably within top three (after USA and China) within this decade. And it is less time consuming to install wind or solar power projects. So, we can expect the balance to shift towards cleaner sources of electricity in the future.

However, electric vehicles have some limitations such as:

- No continuous supply of electricity in India
- No charging points in markets and petrol pumps.
- Govt. will lose tax income on petrol/diesel if these cars get popular in India.
- Expensive electricity.
- General awareness among the public in India about electric cars.
- High maintenance cost of these cars, as it cannot be serviced/ repaired by roadside mechanics.

Electric vehicles are definitely the need of the hour not just due to rising cost of fuel but also due to the fact that natural resources are not expected to last long. Government and manufacturers should think of marketing it as the cheapest mode of transport to attract the consumers.

There are multiple challenges in making it a success for which all stakeholders (government, manufacturers, and the end users) need to work together. This includes (but not limited to) subsidies by the government, making it mandatory for public transport, proper marketing and awareness campaigns, proper disposal of rechargeable battery, solar recharge option (in moving vehicles), and providing battery recharge kiosks.

4.2 Recommendations

Although EV industry has attracted a lot of media attention in recent years along with government focus, there are still not many players who can manufacture the EVs as per consumer demand. The authors provide various recommendations which can be taken up by industry players, government departments and other like-minded people:

4.2.1 Promotion through various institutions:

- The government should promote environmental education through various institutes such as schools, colleges and universities. The youth needs to be sensitized about various side effects of fossil fuels and the need to reduce our dependance on these fuels. This initiative can control environmental damage.
- Through the study it was revealed that media plays an important role in the promotion of any issue, product or concept. The government and the people associated with electric vehicles can develop strong ties with the media, both print and electronic.

4.2.2 Infrastructure development:

- Significant research is required to reduce the charging time of batteries so that people can purchase more electric vehicles in the future. A phased manner approach is required by the government for developing an infrastructure which includes charging stations, better quality of roads and vehicle maintenance related service stations for such vehicles. This will reduce the range anxiety in customers.
- Production of electric three-wheelers is a viable and meaningful option in India. On average, three-wheelers in the metro have a daily run of 120 kilometers, with an average trip length of 8 kilometers, making it very much suitable for an electric prototype. So the automobile manufacturing companies can be immediately motivated with incentives and huge

business opportunities to develop electric three-wheelers with a robust body, designs, and lithium-ion batteries to ensure that they succeed in the market.

- Engineers and automobile manufacturers have to work together to develop a quick charge technique. Recently launched Reva car by Mahindra and Mahindra claims that if the car is charged for 15 minutes, it can travel a distance of 25 kilometers. But for a consumer of a busy city like Delhi, this is a longtime. A consumer in Delhi or Mumbai is so busy that he/she cannot afford to wait for 15 to 30 minutes for a quick charge to travel a distance of 25 kilometers.

4.2.3 Providing initial subsidies:

- Infusion of capital support and government subsidies can play a key role in acquiring new customers and establishing the market for electric vehicles in India. Various subsidies such as exemption from local and state tax, waiving the road tax, exemption from toll taxes and parking charges, access to bus lanes etc can push the demand for such vehicles.
- The Government of India in association with the state governments should encourage and motivate Indian automobile manufacturers through incentives, tax exemption or reduction and other modes to innovate and produce electric vehicles.

4.2.4 Introduce more environmentally friendly vehicles

- Government needs to push the manufacturers to build the electric vehicles as per the needs and expectations of the customers. Greater competition amongst manufacturers will lead to production of cost effective vehicles which will increase the market size and customer base.
- Public transport, specifically buses could be introduced in smaller routes in cities. Electric buses can be utilized as feeder buses.
- There should a good budgetary provision by the government for research & development in electric vehicles and associated areas. Such allocations will motivate the researchers to do extensive research in this area.

4.2.5 Ensure appropriate legislation

- By introducing apolicy and legislation framework, environmental pollution can be controlled. The legislation can include restriction in the uses of private cars or strengthening the public transport system.
- Significant reduction in carbon emissions can be done by electric vehicles, even when using electricity generated primarily from coal; EVs also reduce the use of oil, transmission fluid, and other hazardous fluids and cut noise pollution. The decision makers, therefore, may invite environmentalists, social animators, administrators, members of housing societies etc to support the easy entry of electric vehicles into India through policy reforms.

Chapter 5

Consumer Forums

Consumer forums provide a platform to record the voice of citizens. It also provides critical information which is less accessible through structured questionnaires. The research team organized focus group meetings through consumer forums in Lucknow and New Delhi to understand the consumer preferences for EVs. The objective of the discussions was to understand the consumer preferences, existing constraints of EV and record their opinions regarding various initiatives taken by the Government of India in promoting green mobility.

5.1 Discussion with the stakeholders of Lucknow

This section highlights people's views about electric vehicles, consumer preferences and attitudes, policies related decisions, economic issues and technology concerns.

A focus group meeting of potential stakeholders of electric vehicles in Lucknow with academics, entrepreneurs, car making companies, housewives, socialites, automobile journalists, engineers, policy makers, and social animators – was organized on 25th August 2012 at Hotel Clarks Avadh in Lucknow by the research team and the representative of Swinburne University.

In the introductory note, the research team said that petrol or diesel vehicles are hazardous and polluting whereas EVs are a good alternative. If the EVs are adopted, it will save domestic money which goes outside the country to pay for petrol. Public transport, taxis, autos, school run vehicles, medical vans, etc can be converted to electric vehicles. The team also highlighted that in the West especially in Australia, people have shown their deep interest in adopting electric vehicles as second vehicles. India and China are the two important Asian countries which have great potential for adopting electric vehicles. For this, however, a positive atmosphere, strong awareness, willingness of the technological community, policy makers, social animators, environmentalists and people are needed. Once the awareness about the benefits of EVs is created, making it operational will not be a problem.

According to industry representatives, electric vehicles are already making a mark in the transportation sector in India. Currently, there is only one fully electric vehicle model available in India called Mahindra Reva. For two wheelers there are many players such as Hero electric, YoBikes, TVS etc.

The participants engaged themselves in discussions with an open mind and sense of commitment. The participants shared their experiences and realized the significance of EVs as alternative vehicles in India. Uttar Pradesh being the largest state of India has great potential to adopt EVs. However, some of the participants, mainly the indigenous innovators and small businessmen were a little uncomfortable with government policies, implementation and regulation of policies, the conditions for loan, and the availability of EVs and their parts as well as accessories from the neighboring countries in cheap prices and false claims of watts, efficiency and other things. They do not necessarily provide authentic products; and batteries are also low grade. But this trend does not allow the indigenous makers to flourish in India. Serious steps to check such malpractices are immediately required.

The respondents expressed their opinions and views, as suggested, on the following aspects:

- Technological progress of vehicles and components
- Infrastructure requirements for large-scale adoption
- Economic and long-term economic viability of electric vehicle
- Financing and support mechanisms available
- Regulatory positioning in India and the world
- Government willingness to accept it as public transport
- Subsidy by the government to the buyers and makers of electric vehicles as it is almost pollution free and environment-friendly and, as a result, they are rightly termed as zero-emission cars

- Measures of promotion and popularization and creating public opinion and awareness with the help of electronic and print media
- Prioritizing two-wheelers and autos in tier two cities like Lucknow for experimenting with launch of electric vehicles

Technology

Following were the technology related issues raised in the discussion:

An entrepreneur and owner of Modular Machines from Faridabad who is well connected in Lucknow in connection to meaningful projects initiated the discussion on technology and prospects of EVs in India. He expressed his dissatisfaction with the condition of EVs and rules, regulations, the role of governments, banking institutions and guidelines of the policy makers. As an engineer and innovative entrepreneur, he jumped into the alternative field of energy long ago in 1996 and started making battery driven tri-rickshaws. He followed the guidelines of the government and made the vehicles. But other people who do not follow the guidelines and made false claims did not allow him to prosper. There is no policy implementation; he says in frustration, to check the wrongs in this field. The banks do not approve his loans as according to the guidelines, he has to follow certain formalities. His two college going daughters have opted not to take up a career in automobiles.

The issues that were discussed are:

- Some of the participants felt that the government, the regulatory authorities, makers, and users of electric vehicles in India are in ignorant situations.
- The charging of the batteries of these vehicles is not completely done through solar sources. The battery life is very short and at the end of a battery's life cycle, there is a concern about its environmental impact if lead acid batteries are used.
- According to some of the participants, not everything about EVs is great. The battery of EV is charged via electricity and in India electricity is mostly generated by the thermal stations where the coal is converted into electricity. This conversion of coals into electricity burns a very high percentage and quantum of raw coal to get the electricity. This process causes the emission of carbon dioxide and fuel is also wasted. Therefore

charging through the grid and generation of electricity also has an ecological impact.

- The road conditions in India mainly in rural areas are in bad shape. The road conditions are the major reason for increasing the maintenance costs of motor vehicles. Usually, within the first 6 months of purchase of EV, the vehicle needs to be repaired because of the difficult road conditions, especially the e-rickshaws. The parts of the vehicle need to be changed every year. Maintenance requirements of these vehicles are high due to regular change in spare parts. The parts are also not available easily. In many cases, the consumers buy EVs made in China. They are cheap, low-quality products but attractive. For whatever reason, the price of Chinese products is very low compared to Indian products. This attracts customers to Chinese products. The spare parts of these products have to be imported and thus, we become dependent on another country.
- Electric vehicle market in India is at its nascent stage. Technology must be regulated and awareness needs to be created.
- In order to enhance technology and technology transfer, a time bound mission is needed and policymakers need to be convinced.
- The research potential in terms of scholars and understanding of industrial practices in India is immense. If encouraged and given proper facilities, Indian scientists are capable of developing several alternatives and new technologies in electric vehicles and making them localized as per the demands of Indian roads, people's needs and requirements. The government and industrial companies should allocate more funds for research and development.
- Supporting infrastructure is needed and integration of the laboratory where the research work on these vehicles is done with the development and testing is required.
- Research is not limited to technological development. Special focus on material development should also be there. Increased material research in relation to electric vehicles is required in India.
- Other sources such as solar, wind etc can also be used to generate clean electricity to run the EVs. The government through private and public partnerships must encourage researchers to work on all possible options.

Through coal, we are polluting the atmosphere and also exhausting the limited natural reserves of coal forever. However, if the wind, solar and similar resources are used, they are regenerative and no pollution hazards are associated with them.

5.1.2 Economic

Following were the different economic perspectives of the discussants:

- The government has to play a significant role in the development of the market for EVs in India. Initially, the role of government is required to encourage this market. People are talking about electric vehicles but awareness needs to be created among them for a better understanding.
- Ground level incentives could be given to the consumers of these electric vehicles to promote their usage. Incentives like a rebate or carbon credits could be used for this purpose. Also, the proper implementation of these incentives should be there.
- India is a country which does not have sufficient electric power, so policies need to be made keeping this in mind. But there are limited policy initiatives within states such as Uttar Pradesh.
- A known economic model could be adopted to
    - Create a vehicle which is suitable for the kind of infrastructure available in India
    - A vehicle which can run without any failure
- The funding for electric vehicles in India is very limited. High net worth individuals or corporate organizations are required to support the development of electric vehicle market in India. Venture funds are not actively investing in this field. Though new ideas get funded in Uttar Pradesh, the frequency is not as high as in other industrial areas.
- The small entrepreneurs of EVs should be encouraged by way of easy loans, soft loans, and other incentives to excel in this business and become the role model for others.

- Under the present scheme, REVA and all other EVs are engaged in making low power and low-speed vehicles. If we have to use it as an alternative vehicle, we have to think in terms of making the EVs smart, attractive, high power, high speed, luxurious and friendly.
- In-house manufacturing of electric vehicles including tri-rickshaws, scooters and other agricultural and household implements should be encouraged and promoted in India.
- In many states of India, particularly in Uttar Pradesh, lack of power supply is a common problem. In this situation, the operation of EVs seems impossible. The government has to think seriously about this before launching the EVs.

5.1.3 Policies

The policies which could be adopted and drafted may contain the following points:

- It is envisaged that in the near future, "government subsidies, procurement programs and development of niche markets" will play a critical role in promoting electric vehicle technology.
- Through the proper framing of policy, the government has to indicate that with the introduction of electric vehicles, a constructive shift will take place which will help bring "energy security, air quality, climate change, public health, and economic benefits" to India. This can be achieved by making a research agenda for electric vehicles and developing an action plan.
- The scope of active workshops (open group discussions) should be broadened. Government officials should also be included in these sessions. Groups who could draft detailed reports and plans need to be involved in a platform for encouraging electric vehicle adoption.
- Acommon platform for discussion of ideas and development of environmental technologies is needed.
- There are regulatory malpractices of the Motor Vehicles Act and other associated legislation which need to be addressed.

- The potential of adopting electric vehicles in Uttar Pradesh is high. Adoption of the electric vehicle could be encouraged by making the second vehicle in electric mode mandatory.
- Consumers could be given incentives for using such vehicles. Also, if the cost of the battery could be reduced then users would be able to afford it.

5.1.4 Consumer Preferences and Attitudes

- It is expected that students, mainly girls would prefer to go for EVs only if they are considered safe and cost effective. In a city like Lucknow and Kanpur, the parents do not want their girls to go to the petrol pumps or CNG stations for charging or filling up fuel. Charging their EVs at home with a plug-in system will be easy and comfortable.
- Such vehicles could be used as a second car. Maintenance cost should also be considered.
- Also, the capacity of these vehicles is important. A good brand is also required to make this vehicle successful among people.
- Extra features could be added to make it appealing to get a better response.
- Such vehicles could be used for public or village transportation. A common coach/bus could be brought into the area and people can come and park their vehicles at a place and then they can further travel through these buses/coaches.
- If better replacement options are provided to people, they will definitely consider switching to electric vehicles.
- Conditioning and change in attitude are required through seminars in schools and advertising, to encourage future generations to take up the idea and encourage adoption of green technology.
- Slow adoption of electric vehicles is expected in India.
- With a growing interest amongst consumers, there will be tough competition in the electric vehicle market. This will improve the quality of batteries and will reduce the weight of the vehicle's body which will result in lower prices.

5.1.5 Conclusion

There is a need for a group of people to draw a road map for these zero emission vehicles. Also, a group of technical people should sit and work together on improving super capacitors and drive systems. Marketing and the consumer aspects of the electric vehicle should be focused on.

- Public participation is openly sought. The scientists, environmentalists, leaders, media and the government all have to work together to create awareness among the people of India about green issues and the usefulness of EVs in achieving the target of a cleaner, safer and better ecological atmosphere. The common man must understand the importance of the environment in our lives.
- Consumer confidence needs to be developed over time.
- Under the Public-Private-Partnership (PPP) program, there is an urgent need to develop a treaty between electric companies which are owned by states. The clear role of industry and other administrative departments will contribute to the development of electric vehicles in India.
- The economics of buying is very important. Taxes, incentives and technological additions should be given thorough consideration.
- The field of electric vehicles is growing with a bright future. In order to make this vehicle acceptable amongst the people, policy makers could draft legislations for every household or could take some other regulatory measure to boost the technology and mitigate the impact on climate change.
- There is an urgent need to create an atmosphere where technical solutions should be preferred over behavioral change.
- Cultural dimension can also be focused on while promoting the various policy related amendments.
- Zero emission vehicles should be made cost effective in all possible manners – through regulation and also technological innovation.
- Steps need to be taken to create awareness of zero emission vehicles and the advantages associated with them.
- A sustainable business plan should be developed and the target market for this type of product needs to be identified.

- Electric vehicles could solve the problems of irreversible changes in the environment caused by the activities that emit greenhouse gases. Mitigation of climate change could be done through the adoption of zero emission vehicles.
- A common platform should be provided to the academia to do research work on zero emission vehicles.

5.2 Discussion with the stakeholders of New Delhi[1]

5.2.1 Group discussion at International Institute of Health and Management Research

A stakeholders consultation on the topic: "To identify issues related to consumer preferences and consumer selection for use of electric vehicle" was organized at "International Institute of Health and Management Research at Plot No. 3, HAF Pocket, Sector 18A, Dwarka, Phase- II, New Delhi – 110075" on March 24, 2013 from 4.00-6.00 PM. The purpose of the focus group was to get everyone involved in the conversation and participate actively in the discussion. The participants included academicians, entrepreneurs, technical writers, insurance advisors, automobile journalists, engineers, policy makers, company secretaries and financial advisors.

Through this workshop, the research team of Swinburne University and Amity University, Noida Uttar Pradesh, India tried to understand the opinions and preferences of consumers and various stakeholders of alternative energy about the prospects, problems and future of electric vehicles in India with the help of a carefully structured questionnaire. The team collected data from the consumers and interviewed stakeholders in various cities in India to define:

- Factors which can drive the use of electric vehicles such as "noise pollution and environmental pollution, the cost of petrol and running out of petrol".
- Availability of electric vehicles, including "the purchase price, running cost and servicing cost of such vehicles".

---

[1]Due acknowledgment is made to the researcher on this project at the time, Dr. Kailash Kumar Mishra for his contribution by organizing this forum and documenting the outcomes.

- Feasibility of driving an electric vehicle and additional important factors such as "personal preference, range anxiety, children's influence on environmentally friendly transport, issues with aging population for use of such cars and gender issues".

As discussed during the workshop, electric car adoption can be increased in India if proper attention is given to the following:

1. With a single charge, it can cover a distance of at least 250-300 km (weekly distance traveled by most of the riders)
2. If solar charging option can be used to charge the battery of a running vehicle (solar energy is in abundance), life of a battery can be increased (at least 6-7 years)
3. Preference is given to the owners of electric cars in parking slots, subsidies, toll tax, etc.
4. Increase usage in commercial cars (taxi and cabs), for which point 1 and 2 must be considered

A respondent who is a practicing company secretary, exhibited his concern and curiosity about electric vehicles. He explained his concerns in five broad areas:

1. Technical attention
2. Marketing attention and after sale services
3. Finance to buy or replace the vehicle
4. Cost of production of car to pass on the benefits to customers
5. Establish an action team

He also briefly explained and added sub-points to all the above five issues:

1. Technical attention

1.1. The vehicle requires special services for the battery.

1.2. The research and development initiatives are expected to bring more life to the battery.

1.3. Utilization of technology to generate energy during travel or during parking through wind and solar energy, to add to vehicle life.

2. Marketing attention and after sale services

2.1. Sell vehicles to short distance commuters

2.2. Avail more tax incentives and subsidies to pass on to the customer

2.3. Battery replacement for life, scheme to replace battery at low cost

2.4. Free and compulsory services in the long run with pick up and drop off to facilitate customer satisfaction

2.5. 100% damage claim policy without depreciation through community insurance policy

2.6. Breakdown services with self-team or special agency on a prompt basis

2.7. Spread product knowledge, its working, durability, ability and suitability

3. Finance to buy or replace the vehicle

3.1. Facilitate financing by own finance company or through bankers

3.2. Facility to revalue the vehicle and buyback

4. Cost of production of car to pass on the benefits to customers

4.1. Mass production to reduce the cost

4.2. Bulk procurement to reduce the cost

4.3. Facilitate SMEs for spares to reduce cost, get quality

5. Establish action team

5.1. For government

5.2. For environment

5.3. For the vehicle manufacturers association to reap the benefit of non-polluting vehicles

5.4. To encourage carbon credit utilization

Consensus recommendations

- Awareness drives about the use and importance of electric vehicles in India
- Battery: Electric vehicles manufacturers might provide battery leasing services by which efficiency and cost effectiveness of EVs can be ensured in India.
- Mass mobilization and campaign: EV considerably reduces carbon emissions, even when using electricity generated primarily from coal; reduces oil usage, transmission fluid, and other hazardous fluids; cuts noise pollution. The focus group (FG), therefore, may invite environmentalists, social animators, administrators, members of housing societies etc to support the easy entry of electric vehicles into India.
- Lobbying: Every product gets the people's acceptance and becomes popular after aggressive and well-planned lobbying; the same is true with electric vehicles. The FG has to develop layers of strategies for lobbying to promote the positive elements of electric vehicles.
- The areas where electric vehicles can be used immediately: as public transport in the hospital industry, schools, tourist spots etc. Other areas where EVs can be used are milk and milk product vending, transport of grocery and vegetables, etc.
- Electric vehicles are an easy mode of transport for women and senior citizens: electric vehicles, mainly scooters, are now in many parts of the city used by girls, women and also by senior citizens. They use scooters for commuting locally. They do not have to wait in long queues for getting fuel as it happens at the petrol pump.
- Pressure group: With a creation or formation of a powerful pressure group, the focus group(FG) may convince the common people about the positive attributes of electric vehicles.
- Innovation in electric vehicles is the key factor: The manufacturers and the engineers have to work more systematically to create the following things: aesthetics, space, speed, look, resale value, measures and mechanisms for making them suitable for a long drive, and last but not the least, be an easy driving vehicle.

- Subsidy from the government: The FG should work to create modalities and campaign materials to convince the governments – central as well as state – to encourage the buyers on a mass scale to buy electric vehicles. The best encouragement would be subsidy as in the case of cooking gas, kerosene oil, and other products.
- The technique of using solar energy to recharge the electric vehicle can also be used to increase the driving range of EVs.
- Preference should be given to the owners of electric cars in parking slots, subsidies, toll tax, etc.

5.2.2 Group Discussion at Rukmini Devi Institute of Advanced Studies

A focus group meeting (2nd meeting in Delhi) was organized at Rukmini Devi Institute of Advanced Studies on 18th May 2013. The students and teachers of RDIAS and also the faculty members of other institutions of RDIAS (Rukmini Devi Public School, Pitampura; Rukmini Devi InfoTech, Rohini; Rukmini Devi Lali Kala Kendra, Pitampura; Rukmini Devi Public School, Rohini; Rukmini Devi College of Education, Rohini) participated in the FGM with enthusiasm.

The discussants were ready to discuss, learn, share and voice their opinions on various issues related to electric vehicles in India. It was a unique experience which removed our preconceived notions that the people of India are ignorant about the usefulness, benefits, problems and prospects of electric vehicles. It was a meeting where more than 155 participants actively participated in the discussions and shared their ideas and suggestions. It was a meeting which was attended by young undergraduate students, postgraduate students, teachers, junior, middle and senior faculty members, research scholars from other institutes, media persons from newspapers and All India Radio. Senior citizens, concerned parents and professionals were also the part of the focus group meeting (FGM).

The issues that were discussed are:

- Government willingness to accept it as public transport
- Subsidy by the government to the buyers and makers of electric vehicles as it is almost pollution free and environment-friendly and as a result rightly

termed zero emission cars. The New Delhi government offers a chunky 29.5% subsidy on e20. However, the other state governments are silent. The Government of India in consultation with the state governments should think of giving 35% subsidy for electric vehicles. If it happens, more and more partners will be willing to produce and manufacture electric vehicles.

- Long-term economic value of going for electric vehicles
- Measures of promotion and popularization and creating public opinion and awareness with the help of electronic and print media
- Prioritizing two wheelers: two wheelers are useful for girls, women and senior citizens as they are comfortable, good for a short drive, and one need not to go to petrol pumps for fuel.
- Prioritizing autos for experimenting with electric vehicles.
- Electric vehicles can be a good option for school vans, hospital vans and similar transport systems.
- Marketing and social strategies.

The hall had students, teachers from management and engineering colleges and also teachers and students from Rukmini School.

From the academician group, Director General of RDIAS welcomed the team for choosing RDIAS for the second FGD. In his very brief but meaningful address, he said that in India, fuel prices are directly proportionate to the consumption of fuel. Whenever prices for petrol and diesel are increased, people reduce the use of fuel for their personal vehicles. A few even started using public transport including the metro train, while a few started carpooling. The electric vehicle is a viable and meaningful option in India.

Citing an example of his neighbor, he said: "A friend of mine has been using an electric vehicle. It is very economical compared to petrol-based ones. The electric scooter is easy to handle and there is no need for insurance, tax or even a license for driving these low power electric bikes."

He also suggested that automobile companies like Mahindra and Mahindra should think of putting one electric vehicle at centres like RDIAS so that students can understand easily its utilities, specialties and think of buying, and influencing others to buy electric vehicles.

One professor of RDIAS took the dialogue further and said that the "most important point about an electric bike is its impact on the environment. The electric vehicle operates with zero carbon emission. Most electric vehicles are designed to travel about 80 km with a single charge. It means it will fulfill all your intra-city travel for a day with overnight charging which costs only a fraction of the cost of a petroleum-powered counterpart. Experts found that at least by 2020 main players in the vehicle manufacturing industry would roll out EV models to give confidence to customers. And the industry needs to switch to electric or green powered vehicles in order to result in a considerable reduction in carbon emissions. Leading countries like UK, Brazil and several others have already started operations focused on making electric vehicles popular."

He said: "This is a collective issue for all of us in India. The Indian government should look into this issue seriously, because as a country with the second largest population, India is responsible for lowering the per head carbon footprint to save the earth by promoting non-conventional and green energy sources. There should be action to raise public awareness about journey profiles to help them make informed choices on vehicle requirements."

A student of management studies from the RDIAS reacted to the remark of the professor and said "We, the youth of India have to come forward and convince our people to accept electric vehicles for saving the environment and economy of India. It is, however, commonly observed that the power hungry youth will not choose a slow moving electric vehicle, and the cost, availability of charging points, and service facilities of the electric vehicle will play a major role. We, therefore have to manage everything possible to see that we are creating the perfect atmosphere for electric vehicles in India. An all-round measure is highly needed for such an initiative in India".

According to one academician, researchers in India are seriously engaged in the study of lithium-ion batteries as more viable, meaningful and economical source of alternative energy for electric vehicles in India. One academician highlighted that the role of the research and development wing of IIT Kharagpur is very significant and it has been working hard to develop lithium-ion batteries which will help to run everything from mobile phones to electric cars.

EV Glossary

Below are definitions of many terms commonly used in the report in the context of electric vehicles.

Term	Definitions
AC	Alternating Current
AFV	Alternate Fuel Vehicle — A vehicle powered by fuel other than gasoline or diesel. Examples of alternative fuels are electricity, hydrogen, and CNG.
Ah	Amp Hour — a measure of current (measured in amperes drawn over time (measured in hours), typically used to describe battery capacity and battery charging or discharging.
AT PZEV	Advanced Technology PZEVs — AT PZEVs meet the PZEV requirements and have additional "ZEV-like" characteristics. A dedicated compressed natural gas vehicle, or a hybrid vehicle with engine emissions that meet the PZEV standards would be an AT PZEV.
BEV	Battery Electric Vehicle— An EV powered by electricity stored in batteries.
CARB	California Air Resources Board
CNG	Compressed Natural Gas— Natural gas is primarily methane (CH_4). To serve as a vehicle fuel, it must be stored at high pressure (i.e. compressed) to get sufficient amounts in a tank.
DC	Direct Current
E85	E85 is a blend of 85% ethanol and 15% gasoline by volume.
EV	Electric Vehicle — a vehicle where an electric motor powers the wheels. Battery EVs store electric power in batteries. Hybrid EVs supplement battery storage with electricity generated from another fuel (e.g. gasoline or diesel ICE). Fuel cell EVs supply electricity to the motor from a fuel cell. Additional EAA information on EVs is below.
Fuel Cell	Fuel cells convert fuel directly into electricity. The most common type is a PEM, which converts gaseous hydrogen (H2) into electricity and water (H2O).
FCV	Fuel Cell Vehicle Additional EAA information on FCVs is below.
FFV	Flexible Fuel Vehicle— a vehicle that can run on a variety of fuels (most often just gasoline and E85)
GHG	Green House Gas— A gas that in the atmosphere prevents heat from radiating back into space, and thus warms the earth (thegreenhouse effect). Carbon dioxide (CO_2) is the most common greenhouse gas. Methane (CH_4) is another GHG, and has approximately twenty times the greenhouse effect as CO_2 in the atmosphere.
HEV	Hybrid Electric Vehicle— A vehicle that combines conventional power production (e.g. an ICE) and an electric motor. Additional EAA information on HEVs is below.
HOV	High Occupancy Vehicle
HP	Horsepower— a measure of power (power is energy per time). For engines this is torque multiplied by rotational speed. In the automotive world, it is used to rate engines, but comparisons based on engine horsepower can be misleading because torque varies significantly with RPM for an ICE and the test conditions of the measurement must be carefully specified (e.g. SAE test procedures). In contrast, for electric motors, horsepower is simply defined as 746 Watts (0.746KW).

Source: http://www.electricauto.org/?page=EVGlossary

Term	Definitions
Hz	Hertz— cycles per second, a measure of AC frequency. The U.S. AC electric grid is 60Hz.
ICE	Internal Combustion Engine— An engine that burns fuel inside a reaction chamber to create pressure inside the chamber that is converted into rotary motion. ICE engines are typically based on the Otto cycle, Atkinson cycle, or Wankel engine.
KW	Kilo Watt— a measure of power (power is energy per time).
KWh	Kilo Watt Hour— a measure of energy.
LEV	Low Emission Vehicle — All new cars sold in California starting in 2004 will have at least a LEV or better emissions rating.
LiIon	Lithium Ion— a battery technology consisting of several different chemistries, with different characteristics.
LSV	Low Speed Vehicle — another name for Neighborhood Electric Vehicles rating, not typically permitted on highways.
NEMA	National Electrical Manufacturers Association— Typically used in the EV community when refering to plug and receptacle types, such as ones at NEMA Plug & Receptacle Configurations.
NEV	Neighborhood Electric Vehicle - not typically permitted on highways.
NOx	Nitrogen oxides formed whenever combustion occurs in the presence of nitrogen
NiMH	Nickel Metal Hydride— a battery technology which pre-dates modern Lithium chemistries.
PbA	Lead Acid— a battery technology dating back to the earliest 20th century. .
PEM	Proton Exchange Membrane— a type of fuel cell.
Peukert's Law	Estimates the capacity of an electric battery over a range of discharge rates. See Wikipedia Peukert's law for more information.
PHEV	Plug In Hybrid Electric Vehicle Additional EAA information on PHEVs is below.
PV	Photovoltaic (think rooftop solar)
RPM	Revolutions Per Minute
SAE	Society of Automotive Engineers
SOx	Sulfur oxides
SULEV	Super Ultra Low Emission Vehicle— SULEVs are 90% cleaner than the average new model year car.
ULEV	Ultra Low Emission Vehicle — ULEVs are 50% cleaner than the average new model year car.
ZEV	Zero Emissions Vehicle— ZEVs have zero tailpipe emissions are 98% cleaner than the average new model year vehicle. These include battery electric vehicles and hydrogen fuel cell vehicles.

Source: http://www.electricauto.org/?page=EVGlossary

Annexures

Annexure 1 – Tables

Table 80 : Age Group

Age group of the respondents					
		Frequency	**Percent**	**Valid Percent**	**Cumulative Percent**
Valid	18-30	1267	55.5	56.6	56.6
	31-60	830	36.4	37.1	93.7
	61 and above	141	6.2	6.3	100.0
	Total	2238	98.1	100.0	
Missing	Not Disclosed	43	1.9		
Total		2281	100.0		

Table 81: Gender

Gender					
		Frequency	**Percent**	**Valid Percent**	**Cumulative Percent**
Valid	Male	1273	55.8	56.8	56.8
	Female	970	42.5	43.2	100.0
	Total	2243	98.3	100.0	
Missing	Not Disclosed	38	1.7		
Total		2281	100.0		

Table 82 : Education

Education					
		Frequency	**Percent**	**Valid Percent**	**Cumulative Percent**
Valid	Bachelor's	1056	46.3	47.0	47.0
	Master's	881	38.6	39.2	86.2
	Student	311	13.6	13.8	100.0
	Total	2248	98.6	100.0	
Missing	Not Answered	33	1.4		
Total		2281	100.0		

Table 83 : Monthly income

Monthly income (INR)					
		Frequency	**Percent**	**Valid Percent**	**Cumulative Percent**
Valid	25,000 – 50,000	1216	53.3	59.2	59.2
	50,000 – 100,000	702	30.8	34.2	93.3
	Above 1 lakh	137	6.0	6.7	100.0
	Total	2055	90.1	100.0	
Missing	Not Disclosed	226	9.9		
Total		2281	100.0		

Table 84 : Drive per day

Drive per day					
		Frequency	Percent	Valid Percent	Cumulative Percent
Valid	Up to10 km	348	15.3	16.3	16.3
	11 to 20 km	676	29.6	31.6	47.8
	21 to 30 km	428	18.8	20.0	67.8
	31-60 km	489	21.4	22.8	90.7
	60 km or more	200	8.8	9.3	100.0
	Total	2141	93.9	100.0	
Missing	Not Answered	140	6.1		
Total		2281	100.0		

Table 85 : Awareness source about new technology

Factor	Count	Table N %
Television	1319	58.3%
Newspaper	1339	59.2%
Radio	381	16.9%
Magazines	801	35.4%
Internet	1405	62.2%
Word of mouth	241	10.7%
Other	197	8.7%

Table 86 : Vehicle purchase decision

Who influences your vehicle purchase decision?					
		Frequency	**Percent**	**Valid Percent**	**Cumulative Percent**
Valid	Elders	1162	50.9	52.4	52.4
	Children	287	12.6	12.9	65.3
	Joint decision	769	33.7	34.7	100.0
	Total	2218	97.2	100.0	
Missing	Not Disclosed	63	2.8		
Total		2281	100.0		

Table 87 : About new vehicle

When a new vehicle becomes available for purchase, what do you do?					
		Frequency	**Percent**	**Valid Percent**	**Cumulative Percent**
Valid	I am among the first to purchase it.	321	14.1	14.4	14.4
	I wait to read a review of it and then buy it if the review is favorable	990	43.4	44.4	58.7
	I wait until this new technology has been widely accepted and proven before considering it.	852	37.4	38.2	96.9
	Other	69	3.0	3.1	100.0
	Total	2232	97.9	100.0	
Missing	No Answer	49	2.1		
Total		2281	100.0		

Table 88 : Important factors for choosing the vehicles

	Very Important		Important		Not Important	
Factors	**Count**	**Row N %**	**Count**	**Row N %**	**Count**	**Row N %**
Fuel Efficiency	1876	82.7%	367	16.2%	26	1.1%
Safety	1437	63.4%	724	32.0%	105	4.6%
Vehicle Power	1395	61.8%	774	34.3%	89	3.9%
Purchase Price	1376	60.6%	864	38.1%	30	1.3%
Reliability	1106	48.9%	967	42.7%	189	8.4%
Vehicle Size	1075	47.5%	1049	46.4%	139	6.1%
Expected operating costs	983	43.4%	1209	53.3%	75	3.3%
Reputation of particular vehicle make or model	851	37.6%	1243	54.9%	170	7.5%
Fuel type (e.g. petrol, diesel, CNG)	698	30.9%	1399	61.9%	163	7.2%
Vehicle emissions and pollution	682	30.4%	1373	61.2%	190	8.5%

Table 89 : Seriousness of the problems

	Not a Problem		Problem		Major Problem	
Factors	**Count**	**Row N %**	**Count**	**Row N %**	**Count**	**Row N %**
Importing much of our oil from foreign countries	236	10.5%	1069	47.5%	947	42.1%
Vehicle emissions that contribute to climate change	220	9.7%	1150	51.0%	887	39.3%
Vehicle emissions that affect local air quality	418	18.5%	1031	45.7%	808	35.8%
Unsafe communities because of speeding traffic.	405	18.0%	1095	48.7%	750	33.3%
Traffic noise that you hear at home, work or school	223	9.8%	1439	63.6%	602	26.6%
Traffic congestion that you experience while driving.	673	29.8%	1138	50.4%	447	19.8%

Table 90 : Existing vehicle an electric vehicle

Is your vehicle an electric vehicle?					
		Frequency	Percent	Valid Percent	Cumulative Percent
Valid	Yes	84	3.7	3.8	3.8
	No	2116	92.8	96.2	100.0
	Total	2204	96.6	100.0	
Missing	No Answer	81	3.6		
Total		2281	100.0		

Table 91 : Interested in purchasing electric vehicle

How interested would you be in purchasing an electric motor vehicle once they become easily available in the next couple of years?					
		Frequency	**Percent**	**Valid Percent**	**Cumulative Percent**
Valid	Very interested	650	28.5	28.9	28.9
	Somewhat interested	914	40.1	40.7	69.6
	Not very interested	287	12.6	12.8	82.4
	Not planning to purchase future vehicles	207	9.1	9.2	91.6
	Cannot say	188	8.2	8.4	100.0
	Total	2246	98.5	100.0	
Missing	No Answer	35	1.5		
Total		2281	100.0		

Table 92 : Encouragement of electric vehicles by State and Central Government

Do you think transport policy of the State and Central Government should encourage electric vehicles?					
		Frequency	**Percent**	**Valid Percent**	**Cumulative Percent**
Valid	Yes, but only for public transport	695	30.5	31.1	31.1
	Yes, both for public and private transport	1241	54.4	55.6	86.7
	No, not worth it	296	13.0	13.3	100.0
	Total	2232	97.9	100.0	
Missing	Not Disclosed	49	2.1		
Total		2281	100.0		

Table 93 : Environmental factors

	Not Important		Important		Very Important	
Factors	**Count**	**Row N %**	**Count**	**Row N %**	**Count**	**Row N %**
Driving a vehicle with more advanced or innovative technology	217	9.7%	1271	56.6%	756	33.7%
Reduced dependence on gasoline	545	24.3%	1004	44.7%	696	31.0%
Reduced impact on the environment	185	8.2%	1442	64.2%	618	27.5%
Saving money on the cost of operation (using electricity rather than gasoline)	897	40.0%	986	43.9%	362	16.1%

Table 94 : Vehicle attributes

	Not Important		Important		Very Important	
Factors	**Count**	**Row N %**	**Count**	**Row N %**	**Count**	**Row N %**
On-going maintenance and operative costs (including battery replacement)	129	5.8%	1057	47.5%	1041	46.7%
The ability to carry heavy loads	328	14.7%	962	43.1%	942	42.2%
Reliability of the vehicle	304	13.7%	1009	45.3%	913	41.0%
Availability of desirable vehicle size or style	255	11.4%	1103	49.5%	871	39.1%
Concerns about limited access to plug-in locations	320	14.4%	1074	48.3%	830	37.3%
The need to plug the vehicle in to recharge the battery	202	9.0%	1347	60.3%	684	30.6%
A higher purchase price than for a comparable conventional vehicle	572	25.7%	1265	56.7%	393	17.6%

Table 95 : Views about the environmental impact

	Agree		Disagree		Unsure	
Factors	**Count**	**Row N %**	**Count**	**Row N %**	**Count**	**Row N %**
Cars, minivans, vans, pickups and SUVs are an important source of air pollution anymore.	711	31.9%	1379	61.8%	141	6.3%
Cars, minivans, vans, pickups and SUVs are an important source of the greenhouse gases that many scientists believe are warming the earth's climate	1138	51.1%	650	29.2%	439	19.7%
Exhaust from cars, minivans, vans, pickups and SUVs is an important source of the pollution that causes asthma and makes asthma attacks worse	1021	45.8%	688	30.8%	522	23.4%
Government rules allow minivans, vans, pickups and SUVs to pollute more than passenger cars, for every gallon of gas used.	937	42.2%	919	41.4%	366	16.5%
Government rules require minivans, vans, pickups and SUVs to meet the same miles-per-gallon standards as passenger cars.	772	34.7%	892	40.1%	560	25.2%

Table 96 : Willing to pay a premium for electric vehicle

If very/somewhat interested in electric vehicles would you be willing to pay a premium to purchase an electric vehicle?					
		Frequency	**Percent**	**Valid Percent**	**Cumulative Percent**
Valid	10%	637	27.9	29.4	29.4
	15%	931	40.8	43.0	72.4
	20%	598	26.2	27.6	100.0
	Total	2166	95.0	100.0	
Missing	0% Premium	115	5.0		
Total		2281	100.0		

Annexure 2 – Chi-Square Analysis

Chi-Square Test

Frequencies

Zone			
	Observed N	Expected N	Residual
South India	558	570.3	-12.3
West India	515	570.3	-55.3
North India	1028	570.3	457.8
East India	180	570.3	-390.3
Total	2281		

Age group of the respondents			
	Observed N	Expected N	Residual
18-30	1267	746.0	521.0
31-60	830	746.0	84.0
61 and Above	141	746.0	-605.0
Total	2238		

Gender			
	Observed N	Expected N	Residual
Male	1273	1121.5	151.5
Female	970	1121.5	-151.5
Total	2243		

Education			
	Observed N	Expected N	Residual
Bachelor's	1056	749.3	306.7
Master's	881	749.3	131.7
Student	311	749.3	-438.3
Total	2248		

Drive per day			
	Observed N	Expected N	Residual
Up to10 km	348	428.2	-80.2
11 to 20 km	676	428.2	247.8
21 to 30 km	428	428.2	-.2
31-60 km	489	428.2	60.8
60 km or more	200	428.2	-228.2
Total	2141		

How do you become aware of the new technology - Television			
	Observed N	Expected N	Residual
checked	1319	1130.5	188.5
unchecked	942	1130.5	-188.5
Total	2261		

How do you become aware of the new technology - Newspapers			
	Observed N	Expected N	Residual
checked	1339	1130.0	209.0
unchecked	921	1130.0	-209.0
Total	2260		

How do you become aware of the new technology - Radio			
	Observed N	Expected N	Residual
checked	381	1130.0	-749.0
unchecked	1879	1130.0	749.0
Total	2260		

How do you become aware of the new technology - Magazines			
	Observed N	Expected N	Residual
checked	801	1130.5	-329.5
unchecked	1460	1130.5	329.5
Total	2261		

How do you become aware of the new technology - Internet			
	Observed N	Expected N	Residual
checked	1405	1130.0	275.0
unchecked	855	1130.0	-275.0
Total	2260		

How do you become aware of the new technology - Word of mouth			
	Observed N	Expected N	Residual
checked	241	1130.0	-889.0
unchecked	2019	1130.0	889.0
Total	2260		

How do you become aware of the new technology - Other			
	Observed N	Expected N	Residual
checked	197	1129.0	-932.0
unchecked	2061	1129.0	932.0
Total	2258		

Who influences your vehicle purchase decision?			
	Observed N	Expected N	Residual
Elders	1162	739.3	422.7
Children	287	739.3	-452.3
Joint decision	769	739.3	29.7
Total	2218		

When a new vehicle becomes available for purchase, what do you do?			
	Observed N	Expected N	Residual
I am among the first to purchase it.	321	558.0	-237.0
I wait to read a review of it and then buy it if the review is favorable	990	558.0	432.0
I wait until this new technology has been widely accepted and proven before considering it.	852	558.0	294.0
Other	69	558.0	-489.0
Total	2232		

Fuel efficiency			
	Observed N	Expected N	Residual
Very Important	1876	756.3	1119.7
Important	367	756.3	-389.3
Not Important	26	756.3	-730.3
Total	2269		

Safety			
	Observed N	Expected N	Residual
Very Important	1376	756.7	619.3
Important	864	756.7	107.3
Not Important	30	756.7	-726.7
Total	2270		

Vehicle power			
	Observed N	Expected N	Residual
Very Important	1437	755.3	681.7
Important	724	755.3	-31.3
Not Important	105	755.3	-650.3
Total	2266		

Purchase price			
	Observed N	Expected N	Residual
Very Important	983	755.7	227.3
Important	1209	755.7	453.3
Not Important	75	755.7	-680.7
Total	2267		

Reliability			
	Observed N	Expected N	Residual
Very Important	1395	752.7	642.3
Important	774	752.7	21.3
Not Important	89	752.7	-663.7
Total	2258		

Vehicle size (to accommodate passengers or cargo)			
	Observed N	Expected N	Residual
Very Important	698	753.3	-55.3
Important	1399	753.3	645.7
Not Important	163	753.3	-590.3
Total	2260		

Expected operating costs (for maintenance and repair)			
	Observed N	Expected N	Residual
Very Important	1075	754.3	320.7
Important	1049	754.3	294.7
Not Important	139	754.3	-615.3
Total	2263		

Reputation of particular vehicle make or model			
	Observed N	Expected N	Residual
Very Important	682	748.3	-66.3
Important	1373	748.3	624.7
Not Important	190	748.3	-558.3
Total	2245		

Fuel type (e.g. petrol, diesel, CNG)			
	Observed N	Expected N	Residual
Very Important	1106	754.0	352.0
Important	967	754.0	213.0
Not Important	189	754.0	-565.0
Total	2262		

Vehicle emissions and pollution			
	Observed N	Expected N	Residual
Very Important	851	754.7	96.3
Important	1243	754.7	488.3
Not Important	170	754.7	-584.7
Total	2264		

Traffic congestion that you experience while driving			
	Observed N	Expected N	Residual
Not a Problem	673	752.7	-79.7
Problem	1138	752.7	385.3
Major Problem	447	752.7	-305.7
Total	2258		

Traffic noise that you hear at home, work or school			
	Observed N	Expected N	Residual
Not a Problem	223	754.7	-531.7
Problem	1439	754.7	684.3
Major Problem	602	754.7	-152.7
Total	2264		

Vehicle emissions that affect local air quality			
	Observed N	Expected N	Residual
Not a Problem	418	752.3	-334.3
Problem	1031	752.3	278.7
Major Problem	808	752.3	55.7
Total	2257		

Vehicle emissions that contribute to climate change			
	Observed N	Expected N	Residual
Not a Problem	220	752.3	-532.3
Problem	1150	752.3	397.7
Major Problem	887	752.3	134.7
Total	2257		

Unsafe communities because of speeding traffic			
	Observed N	Expected N	Residual
Not a Problem	405	750.0	-345.0
Problem	1095	750.0	345.0
Major Problem	750	750.0	.0
Total	2250		

Importing much of our oil from foreign countries			
	Observed N	Expected N	Residual
Not a Problem	236	750.7	-514.7
Problem	1069	750.7	318.3
Major Problem	947	750.7	196.3
Total	2252		

Is your vehicle an electric vehicle?			
	Observed N	Expected N	Residual
Yes	84	551.0	-467.0
No	2116	551.0	1565.0
Total	2200		

How interested would you be in purchasing an electric motor vehicle once they become easily available in the next couple of years?			
	Observed N	Expected N	Residual
Very interested	650	449.2	200.8
Somewhat interested	914	449.2	464.8
Not very interested	287	449.2	-162.2
Not planning to purchase future vehicles	207	449.2	-242.2
Cannot say	188	449.2	-261.2
Total	2246		

Do you think transport policy of the State and Central Government should encourage electric vehicles?			
	Observed N	Expected N	Residual
Yes, but only for public transport	695	744.0	-49.0
Yes, both for public and private transport	1241	744.0	497.0
No, not worth it	296	744.0	-448.0
Total	2232		

Saving money on the cost of operation (using electricity rather than gasoline)			
	Observed N	Expected N	Residual
Not Important	897	748.3	148.7
Important	986	748.3	237.7
Very Important	362	748.3	-386.3
Total	2245		

Reduced impact on the environment			
	Observed N	Expected N	Residual
Not Important	185	748.3	-563.3
Important	1442	748.3	693.7
Very Important	618	748.3	-130.3
Total	2245		

Reduced dependence on gasoline			
	Observed N	Expected N	Residual
Not Important	545	748.3	-203.3
Important	1004	748.3	255.7
Very Important	696	748.3	-52.3
Total	2245		

Driving a vehicle with more advanced or innovative technology			
	Observed N	Expected N	Residual
Not Important	217	748.0	-531.0
Important	1271	748.0	523.0
Very Important	756	748.0	8.0
Total	2244		

A higher purchase price than for a comparable conventional vehicle			
	Observed N	Expected N	Residual
Not Important	572	743.3	-171.3
Important	1265	743.3	521.7
Very Important	393	743.3	-350.3
Total	2230		

The need to plug the vehicle in to recharge the battery			
	Observed N	Expected N	Residual
Not Important	202	744.3	-542.3
Important	1347	744.3	602.7
Very Important	684	744.3	-60.3
Total	2233		

Concerns about limited access to plug-in locations			
	Observed N	Expected N	Residual
Not Important	320	741.3	-421.3
Important	1074	741.3	332.7
Very Important	830	741.3	88.7
Total	2224		

Availability of desirable vehicle size or style			
	Observed N	Expected N	Residual
Not Important	255	743.0	-488.0
Important	1103	743.0	360.0
Very Important	871	743.0	128.0
Total	2229		

Reliability of the vehicle			
	Observed N	Expected N	Residual
Not Important	304	742.0	-438.0
Important	1009	742.0	267.0
Very Important	913	742.0	171.0
Total	2226		

On-going maintenance and operative costs (including battery replacement)			
	Observed N	Expected N	Residual
Not Important	129	742.3	-613.3
Important	1057	742.3	314.7
Very Important	1041	742.3	298.7
Total	2227		

The ability to carry heavy loads			
	Observed N	Expected N	Residual
Not Important	328	744.0	-416.0
Important	962	744.0	218.0
Very Important	942	744.0	198.0
Total	2232		

Cars, minivans, vans, pickups and SUVs are not an important source of air pollution anymore			
	Observed N	Expected N	Residual
Agree	1379	743.7	635.3
Disagree	711	743.7	-32.7
Unsure	141	743.7	-602.7
Total	2231		

Government rules allow minivans, vans, pickups and SUVs to pollute more than passenger cars for every gallon of gas used			
	Observed N	Expected N	Residual
Agree	937	740.7	196.3
Disagree	919	740.7	178.3
Unsure	366	740.7	-374.7
Total	2222		

Cars, minivans, vans, pickups and SUVs are an important source of the greenhouse gases that many scientists believe are warming the earth's climate			
	Observed N	Expected N	Residual
Agree	1138	742.3	395.7
Disagree	650	742.3	-92.3
Unsure	439	742.3	-303.3
Total	2227		

Government rules require minivans, vans, pickups and SUVs to meet the same miles-per-gallon standards as passenger cars			
	Observed N	Expected N	Residual
Agree	772	741.3	30.7
Disagree	892	741.3	150.7
Unsure	560	741.3	-181.3
Total	2224		

Exhaust from cars, minivans, vans, pickups and SUVs is an important source of the pollution that causes asthma and makes asthma attacks worse			
	Observed N	Expected N	Residual
Agree	1021	743.7	277.3
Disagree	688	743.7	-55.7
Unsure	522	743.7	-221.7
Total	2231		

If very/somewhat interested in electric vehicles would you be willing to pay a premium to purchase an electric vehicle?			
	Observed N	Expected N	Residual
10%	637	722.0	-85.0
15%	931	722.0	209.0
20%	598	722.0	-124.0
Total	2166		

Annexure 3 – Chi-Square Summary

Variables	Chi-Square	df	Asymp. Sig.
Zone	640.128[a]	3	.000
Gender	40.931[b]	1	.000
Education	405.049[c]	2	.000
Monthly income (INR)	850.444[f]	2	.000
Drive per day	288.671[l]	4	.000
How do you become aware of the new technology – Television	62.861[m]	1	.000
How do you become aware of the new technology – Newspapers	77.312[n]	1	.000
How do you become aware of the new technology – Radio	992.922[n]	1	.000
How do you become aware of the new technology – Magazines	192.075[m]	1	.000
How do you become aware of the new technology – Internet	133.850[n]	1	.000
How do you become aware of the new technology - Word of mouth	1398.798[n]	1	.000
How do you become aware of the new technology - Other	1538.749[o]	1	0.000
Who influences your vehicle purchase decision?	519.566[p]	2	.000
When a new vehicle becomes available for purchase, what do you do?	1018.548[q]	3	.000
Fuel efficiency	2563.183[r]	2	0.000
Safety	1220.007[s]	2	.000
Vehicle power	1176.414[t]	2	.000
Purchase price	2022.685[u]	3	0.000
Reliability	1133.969[v]	2	.000
Vehicle size (to accommodate passengers or cargo)	1020.054[w]	2	.000
Expected operating costs (for maintenance and repair)	753.368[x]	2	.000
Reputation of particular vehicle make or model	943.890[y]	2	.000
Fuel type (e.g. petrol, diesel, CNG)	647.875[z]	2	.000
Vehicle emissions and pollution	781.252[aa]	2	.000

Traffic congestion that you experience while driving	329.841[v]	2	.000
Traffic noise that you hear at home, work or school	1026.001[aa]	2	.000
Vehicle emissions that affect local air quality	255.914[ab]	2	.000
Vehicle emissions that contribute to climate change	610.969[ab]	2	.000
Unsafe communities because of speeding traffic	317.400[ac]	2	.000
Importing much of our oil from foreign countries	539.207[ad]	2	.000
Is your vehicle an electric vehicle?	5934.878[ae]	3	0.000
How interested would you be in purchasing an electric motor vehicle once they become easily available in the next couple of years?	911.743[af]	4	.000
Do you think transport policy of the State and Central Government should encourage electric vehicles?	604.992[ag]	2	.000
Saving money on the cost of operation (using electricity rather than gasoline)	304.464[y]	2	.000
Reduced impact on the environment	1089.761[y]	2	.000
Reduced dependence on gasoline	146.257[y]	2	.000
Driving a vehicle with more advanced or innovative technology	742.719[ah]	2	.000
Making a personal statement	52.161[ai]	2	.000
A higher purchase price than for a comparable conventional vehicle	570.706[aj]	2	.000
The need to plug the vehicle in to recharge the battery	888.006[ak]	2	.000
Concerns about limited access to plug-in locations	399.349[al]	2	.000
Availability of desirable vehicle size or style	516.996[am]	2	.000
Reliability of the vehicle	394.035[an]	2	.000
On-going maintenance and operative costs (including battery replacement)	760.298[ao]	2	.000
The ability to carry heavy loads	349.172[ag]	2	.000
Cars, minivans, vans, pickups and SUVs are not an important source of air pollution anymore.	1032.617[ap]	2	.000
Government rules allow minivans, vans, pickups and SUVs to pollute more than passenger cars, for every gallon of gas used.	284.507[aq]	2	.000
Cars, minivans, vans, pickups and SUVs are an important source of the greenhouse gases that many scientists believe are warming the earth's climate	346.325[ao]	2	.000

Government rules require minivans, vans, pickups and SUVs to meet the same miles-per-gallon standards as passenger cars	76.245[al]	2	.000
Exhaust from cars, minivans, vans, pickups and SUVs is an important source of the pollution that causes asthma and makes asthma attacks worse	173.665[ap]	2	.000
If very/somewhat interested in electric vehicles would you be willing to pay a premium to purchase an **electric** vehicle?	91.803[ar]	2	.000
Age group of the respondents	863.971[as]	2	.000
a. 0 cells (0.0%) have expected frequencies less than 5. The minimum expected cell frequency is 570.3.			
b. 0 cells (0.0%) have expected frequencies less than 5. The minimum expected cell frequency is 1121.5.			
c. 0 cells (0.0%) have expected frequencies less than 5. The minimum expected cell frequency is 749.3.			
d. 0 cells (0.0%) have expected frequencies less than 5. The minimum expected cell frequency is 49.3.			
e. 0 cells (0.0%) have expected frequencies less than 5. The minimum expected cell frequency is 90.3.			
f. 0 cells (0.0%) have expected frequencies less than 5. The minimum expected cell frequency is 685.0.			
g. 0 cells (0.0%) have expected frequencies less than 5. The minimum expected cell frequency is 211.7.			
h. 0 cells (0.0%) have expected frequencies less than 5. The minimum expected cell frequency is 68.0.			
i. 0 cells (0.0%) have expected frequencies less than 5. The minimum expected cell frequency is 94.3.			
j. 0 cells (0.0%) have expected frequencies less than 5. The minimum expected cell frequency is 178.0.			
k. 0 cells (0.0%) have expected frequencies less than 5. The minimum expected cell frequency is 80.0.			
l. 0 cells (0.0%) have expected frequencies less than 5. The minimum expected cell frequency is 428.2.			
m. 0 cells (0.0%) have expected frequencies less than 5. The minimum expected cell frequency is 1130.5.			
n. 0 cells (0.0%) have expected frequencies less than 5. The minimum expected cell frequency is 1130.0.			
o. 0 cells (0.0%) have expected frequencies less than 5. The minimum expected cell frequency is 1129.0.			
p. 0 cells (0.0%) have expected frequencies less than 5. The minimum expected cell frequency is 739.3.			
q. 0 cells (0.0%) have expected frequencies less than 5. The minimum expected cell frequency is 558.0.			

r. 0 cells (0.0%) have expected frequencies less than 5. The minimum expected cell frequency is 756.3.
s. 0 cells (0.0%) have expected frequencies less than 5. The minimum expected cell frequency is 756.7.
t. 0 cells (0.0%) have expected frequencies less than 5. The minimum expected cell frequency is 755.3.
u. 0 cells (0.0%) have expected frequencies less than 5. The minimum expected cell frequency is 566.8.
v. 0 cells (0.0%) have expected frequencies less than 5. The minimum expected cell frequency is 752.7.
w. 0 cells (0.0%) have expected frequencies less than 5. The minimum expected cell frequency is 753.3.
x. 0 cells (0.0%) have expected frequencies less than 5. The minimum expected cell frequency is 754.3.
y. 0 cells (0.0%) have expected frequencies less than 5. The minimum expected cell frequency is 748.3.
z. 0 cells (0.0%) have expected frequencies less than 5. The minimum expected cell frequency is 754.0.
aa. 0 cells (0.0%) have expected frequencies less than 5. The minimum expected cell frequency is 754.7.
ab. 0 cells (0.0%) have expected frequencies less than 5. The minimum expected cell frequency is 752.3.
ac. 0 cells (0.0%) have expected frequencies less than 5. The minimum expected cell frequency is 750.0.
ad. 0 cells (0.0%) have expected frequencies less than 5. The minimum expected cell frequency is 750.7.
ae. 0 cells (0.0%) have expected frequencies less than 5. The minimum expected cell frequency is 551.0.
af. 0 cells (0.0%) have expected frequencies less than 5. The minimum expected cell frequency is 449.2.
ag. 0 cells (0.0%) have expected frequencies less than 5. The minimum expected cell frequency is 744.0.
ah. 0 cells (0.0%) have expected frequencies less than 5. The minimum expected cell frequency is 748.0.
ai. 0 cells (0.0%) have expected frequencies less than 5. The minimum expected cell frequency is 732.7.
aj. 0 cells (0.0%) have expected frequencies less than 5. The minimum expected cell frequency is 743.3.
ak. 0 cells (0.0%) have expected frequencies less than 5. The minimum expected cell frequency is 744.3.
al. 0 cells (0.0%) have expected frequencies less than 5. The minimum expected cell frequency is 741.3.
am. 0 cells (0.0%) have expected frequencies less than 5. The minimum expected cell frequency is 743.0.

an. 0 cells (0.0%) have expected frequencies less than 5. The minimum expected cell frequency is 742.0.
ao. 0 cells (0.0%) have expected frequencies less than 5. The minimum expected cell frequency is 742.3.
ap. 0 cells (0.0%) have expected frequencies less than 5. The minimum expected cell frequency is 743.7.
aq. 0 cells (0.0%) have expected frequencies less than 5. The minimum expected cell frequency is 740.7.
ar. 0 cells (0.0%) have expected frequencies less than 5. The minimum expected cell frequency is 722.0.
as. 0 cells (0.0%) have expected frequencies less than 5. The minimum expected cell frequency is 746.0.

Annexure 4- Hypothesis Test Summary

Hypothesis Test Summary

	Null Hypothesis	Test	Sig.	Decision
1	The categories defined by Q3: Gender = Male and Female occur with probabilities 0.5 and 0.5.	One-Sample Binomial Test	.000	Reject the null hypothesis.
2	The categories of Q4: Education occur with equal probabilities.	One-Sample Chi-Square Test	.000	Reject the null hypothesis.
3	The categories of Q6: Monthly income occur with equal probabilities.	One-Sample Chi-Square Test	.000	Reject the null hypothesis.
4	The categories of Q12: Drive per day occur with equal probabilities.	One-Sample Chi-Square Test	.000	Reject the null hypothesis.
5	The categories defined by Q13a: How do you become aware of the new technology - Television = checked and unchecked occur with probabilities 0.5 and 0.5.	One-Sample Binomial Test	.000	Reject the null hypothesis.
6	The categories defined by Q13b: How do you become aware of the new technology - Newspaper = checked and unchecked occur with probabilities 0.5 and 0.5.	One-Sample Binomial Test	.000	Reject the null hypothesis.
7	The categories defined by Q13c: How do you become aware of the new technology - Radio = unchecked and checked occur with probabilities 0.5 and 0.5.	One-Sample Binomial Test	.000	Reject the null hypothesis.
8	The categories defined by Q13d: How do you become aware of the new technology - Magazines = checked and unchecked occur with probabilities 0.5 and 0.5.	One-Sample Binomial Test	.000	Reject the null hypothesis.
9	The categories defined by Q13e: How do you become aware of the new technology - Computer = unchecked and checked occur with probabilities 0.5 and 0.5.	One-Sample Binomial Test	.000	Reject the null hypothesis.
10	The categories defined by Q13f: How do you become aware of the new technology - Word of mouth = checked and unchecked occur with probabilities 0.5 and 0.5.	One-Sample Binomial Test	.000	Reject the null hypothesis.

Asymptotic significances are displayed. The significance level is .05.

Hypothesis Test Summary

	Null Hypothesis	Test	Sig.	Decision
11	The categories defined by Q13g: How do you become aware of the new technology - Other = unchecked and checked occur with probabilities 0.5 and 0.5.	One-Sample Binomial Test	.000	Reject the null hypothesis.
12	The categories of Q14: Who influences your vehicle purchase decision occur with equal probabilities.	One-Sample Chi-Square Test	.000	Reject the null hypothesis.
13	The categories of Q15: When a new vehicle becomes available for purchase, what do you do? occur with equal probabilities.	One-Sample Chi-Square Test	.000	Reject the null hypothesis.
14	The categories of Q16a:Fuel efficiency occur with equal probabilities.	One-Sample Chi-Square Test	.000	Reject the null hypothesis.
15	The categories of Q16b: Safety occur with equal probabilities.	One-Sample Chi-Square Test	.000	Reject the null hypothesis.
16	The categories of Q16c: Vehicle power occur with equal probabilities.	One-Sample Chi-Square Test	.000	Reject the null hypothesis.
17	The categories of Q16d:Purchase price occur with equal probabilities	One-Sample Chi-Square Test	.000	Reject the null hypothesis.
18	The categories of Q16e:Reliability occur with equal probabilities.	One-Sample Chi-Square Test	.000	Reject the null hypothesis.
19	The categories of Q16f: Vehicle size (to accommodate passengers or cargo) occur with equal probabilities.	One-Sample Chi-Square Test	.000	Reject the null hypothesis.
20	The categories of Q16g:Expected operating costs (for maintenance and repair) occur with equal probabilities.	One-Sample Chi-Square Test	.000	Reject the null hypothesis.

Asymptotic significances are displayed. The significance level is .05.

Hypothesis Test Summary

	Null Hypothesis	Test	Sig.	Decision
21	The categories of Q16h:Reputation of particular vehicle make or mode occur with equal probabilities.	One-Sample Chi-Square Test	.000	Reject the null hypothesis.
22	The categories of Q16i:Fuel type (e g. petrol, diesel, CNG) occur with equal probabilities.	One-Sample Chi-Square Test	.000	Reject the null hypothesis.
23	The categories of Q16j:Vehicle emissions and pollution occur with equal probabilities.	One-Sample Chi-Square Test	.000	Reject the null hypothesis.
24	The categories of Q17a: Traffic congestion that you experience while driving. occur with equal probabilities.	One-Sample Chi-Square Test	.000	Reject the null hypothesis.
25	The categories of Q17b:Traffic noise that you hear at home, work, or school occur with equal probabilities.	One-Sample Chi-Square Test	.000	Reject the null hypothesis.
26	The categories of Q17c:Vehicle emissions that affect local air quality occur with equal probabilities.	One-Sample Chi-Square Test	.000	Reject the null hypothesis.
27	The categories of Q17d:Vehicle emissions that contribute to global climate change occur with equal probabilities.	One-Sample Chi-Square Test	.000	Reject the null hypothesis.
28	The categories of Q17e: Unsafe communities because of speeding traffic. occur with equal probabilities.	One-Sample Chi-Square Test	.000	Reject the null hypothesis.
29	The categories of Q17f: Importing much of our oil from foreign countries occur with equal probabilities.	One-Sample Chi-Square Test	.000	Reject the null hypothesis.
30	The categories of Q18: Is your vehicle an electric vehicle occur with equal probabilities.	One-Sample Chi-Square Test	.000	Reject the null hypothesis.

Asymptotic significances are displayed. The significance level is .05.

Hypothesis Test Summary

	Null Hypothesis	Test	Sig.	Decision
31	The categories of Q20: How interested would you be in purchasing an electric motor vehicle once they become easily available in the next couple of years? occur with equal probabilities.	One-Sample Chi-Square Test	.000	Reject the null hypothesis.
32	The categories of Q21: Do you think transport policy of the state and Central Government should encourage electric vehicles? occur with equal probabilities.	One-Sample Chi-Square Test	.000	Reject the null hypothesis.
33	The categories of Q22a:Saving money on the cost of operation (using electricity rather than gasoline) occur with equal probabilities.	One-Sample Chi-Square Test	.000	Reject the null hypothesis.
34	The categories of Q22b:Reduced impact on the environment occur with equal probabilities.	One-Sample Chi-Square Test	.000	Reject the null hypothesis.
35	The categories of Q22c: Reduced dependence on gasoline occur with equal probabilities.	One-Sample Chi-Square Test	.000	Reject the null hypothesis.
36	The categories of Q22d: Driving a vehicle with more advanced or innovative technology occur with equal probabilities.	One-Sample Chi-Square Test	.000	Reject the null hypothesis.
37	The categories of Q23a: A higher purchase price than for a comparable conventional vehicle occur with equal probabilities.	One-Sample Chi-Square Test	.000	Reject the null hypothesis.
38	The categories of Q23b: The need to plug the vehicle in to recharge the battery occur with equal probabilities.	One-Sample Chi-Square Test	.000	Reject the null hypothesis.
39	The categories of Q23c:Concerns about limited access to plug-in locations occur with equal probabilities.	One-Sample Chi-Square Test	.000	Reject the null hypothesis.

Asymptotic significances are displayed. The significance level is .05.

Hypothesis Test Summary

	Null Hypothesis	Test	Sig.	Decision
40	The categories of Q23d:Availability of desirable vehicle size or style occur with equal probabilities.	One-Sample Chi-Square Test	.000	Reject the null hypothesis.
41	The categories of Q23e:Reliability of the vehicle occur with equal probabilities.	One-Sample Chi-Square Test	.000	Reject the null hypothesis.
42	The categories of Q23f:On-going maintenance and operative costs (including battery replacement) occur with equal probabilities.	One-Sample Chi-Square Test	.000	Reject the null hypothesis.
43	The categories of Q23g: The ability to carry heavy loads occur with equal probabilities.	One-Sample Chi-Square Test	.000	Reject the null hypothesis.
44	The categories of Q24a: Cars, minivans, vans, pickups, and SUVs are not an important source of air pollution anymore. occur with equal probabilities.	One-Sample Chi-Square Test	.000	Reject the null hypothesis.
45	The categories of Q24b: Government rules allow minivans, vans, pickups, and SUVs to pollute more than passenger cars, for every gallon of gas used. occur with equal probabilities.	One-Sample Chi-Square Test	.000	Reject the null hypothesis.
46	The categories of Q24c:Cars, minivans, vans, pickups, and SUVs are an important source of the greenhouse gases that many scientists believe are warming the earth's climate occur with equal probabilities.	One-Sample Chi-Square Test	.000	Reject the null hypothesis.
47	The categories of Q24d: Government rules require minivans, vans, pickups, and SUVs to meet the same miles-per-gallon standards as passenger cars. occur with equal probabilities.	One-Sample Chi-Square Test	.000	Reject the null hypothesis.

Asymptotic significances are displayed. The significance level is .05.

Hypothesis Test Summary

	Null Hypothesis	Test	Sig.	Decision
1	The categories defined by Q3: Gender = Male and Female occur with probabilities 0.5 and 0.5.	One-Sample Binomial Test	.000	Reject the null hypothesis.
2	The categories of Q4: Education occur with equal probabilities.	One-Sample Chi-Square Test	.000	Reject the null hypothesis.
3	The categories of Q6: Monthly income occur with equal probabilities.	One-Sample Chi-Square Test	.000	Reject the null hypothesis.
4	The categories of Q12: Drive per day occur with equal probabilities.	One-Sample Chi-Square Test	.000	Reject the null hypothesis.
5	The categories defined by Q13a: How do you become aware of the new technology - Television = checked and unchecked occur with probabilities 0.5 and 0.5.	One-Sample Binomial Test	.000	Reject the null hypothesis.
6	The categories defined by Q13b: How do you become aware of the new technology - Newspaper = checked and unchecked occur with probabilities 0.5 and 0.5.	One-Sample Binomial Test	.000	Reject the null hypothesis.
7	The categories defined by Q13c: How do you become aware of the new technology - Radio = unchecked and checked occur with probabilities 0.5 and 0.5.	One-Sample Binomial Test	.000	Reject the null hypothesis.
8	The categories defined by Q13d: How do you become aware of the new technology - Magazines = checked and unchecked occur with probabilities 0.5 and 0.5.	One-Sample Binomial Test	.000	Reject the null hypothesis.
9	The categories defined by Q13e: How do you become aware of the new technology - Computer = unchecked and checked occur with probabilities 0.5 and 0.5.	One-Sample Binomial Test	.000	Reject the null hypothesis.
10	The categories defined by Q13f: How do you become aware of the new technology - Word of mouth = checked and unchecked occur with probabilities 0.5 and 0.5.	One-Sample Binomial Test	.000	Reject the null hypothesis.
11	The categories defined by Q13g: How do you become aware of the new technology - Other = unchecked and checked occur with probabilities 0.5 and 0.5.	One-Sample Binomial Test	.000	Reject the null hypothesis.

Asymptotic significances are displayed. The significance level is .05.

Hypothesis Test Summary

	Null Hypothesis	Test	Sig.	Decision
12	The categories of Q14: Who influences your vehicle purchase decision occur with equal probabilities.	One-Sample Chi-Square Test	.000	Reject the null hypothesis.
13	The categories of Q15: When a new vehicle becomes available for purchase, what do you do? occur with equal probabilities.	One-Sample Chi-Square Test	.000	Reject the null hypothesis.
14	The categories of Q16a:Fuel efficiency occur with equal probabilities.	One-Sample Chi-Square Test	.000	Reject the null hypothesis.
15	The categories of Q16b: Safety occur with equal probabilities.	One-Sample Chi-Square Test	.000	Reject the null hypothesis.
16	The categories of Q16c: Vehicle power occur with equal probabilities.	One-Sample Chi-Square Test	.000	Reject the null hypothesis.
17	The categories of Q16d:Purchase price occur with equal probabilities.	One-Sample Chi-Square Test	.000	Reject the null hypothesis.
18	The categories of Q16e:Reliability occur with equal probabilities.	One-Sample Chi-Square Test	.000	Reject the null hypothesis.
19	The categories of Q16f: Vehicle size (to accommodate passengers or cargo) occur with equal probabilities.	One-Sample Chi-Square Test	.000	Reject the null hypothesis.
20	The categories of Q16g:Expected operating costs (for maintenance and repair) occur with equal probabilities.	One-Sample Chi-Square Test	.000	Reject the null hypothesis.
21	The categories of Q16h:Reputation of particular vehicle make or mode occur with equal probabilities.	One-Sample Chi-Square Test	.000	Reject the null hypothesis.
22	The categories of Q16i:Fuel type (e g. petrol, diesel, CNG) occur with equal probabilities.	One-Sample Chi-Square Test	.000	Reject the null hypothesis.
23	The categories of Q16j:Vehicle emissions and pollution occur with equal probabilities.	One-Sample Chi-Square Test	.000	Reject the null hypothesis.

Asymptotic significances are displayed. The significance level is .05.

Hypothesis Test Summary

	Null Hypothesis	Test	Sig.	Decision
24	The categories of Q17a: Traffic congestion that you experience while driving. occur with equal probabilities.	One-Sample Chi-Square Test	.000	Reject the null hypothesis.
25	The categories of Q17b:Traffic noise that you hear at home, work, or school occur with equal probabilities.	One-Sample Chi-Square Test	.000	Reject the null hypothesis.
26	The categories of Q17c:Vehicle emissions that affect local air quality occur with equal probabilities.	One-Sample Chi-Square Test	.000	Reject the null hypothesis.
27	The categories of Q17d:Vehicle emissions that contribute to global climate change occur with equal probabilities.	One-Sample Chi-Square Test	.000	Reject the null hypothesis.
28	The categories of Q17e: Unsafe communities because of speeding traffic. occur with equal probabilities.	One-Sample Chi-Square Test	.000	Reject the null hypothesis.
29	The categories of Q17f: Importing much of our oil from foreign countries occur with equal probabilities.	One-Sample Chi-Square Test	.000	Reject the null hypothesis.
30	The categories of Q18: Is your vehicle an electric vehicle occur with equal probabilities.	One-Sample Chi-Square Test	.000	Reject the null hypothesis.
31	The categories of Q20: How interested would you be in purchasing an electric motor vehicle once they become easily available in the next couple of years? occur with equal probabilities.	One-Sample Chi-Square Test	.000	Reject the null hypothesis.
32	The categories of Q21: Do you think transport policy of the state and Central Government should encourage electric vehicles? occur with equal probabilities.	One-Sample Chi-Square Test	.000	Reject the null hypothesis.
33	The categories of Q22a:Saving money on the cost of operation (using electricity rather than gasoline) occur with equal probabilities.	One-Sample Chi-Square Test	.000	Reject the null hypothesis.

Asymptotic significances are displayed. The significance level is .05.

Hypothesis Test Summary

	Null Hypothesis	Test	Sig.	Decision
34	The categories of Q22b:Reduced impact on the environment occur with equal probabilities.	One-Sample Chi-Square Test	.000	Reject the null hypothesis.
35	The categories of Q22c: Reduced dependence on gasoline occur with equal probabilities.	One-Sample Chi-Square Test	.000	Reject the null hypothesis.
36	The categories of Q22d: Driving a vehicle with more advanced or innovative technology occur with equal probabilities.	One-Sample Chi-Square Test	.000	Reject the null hypothesis.
37	The categories of Q23a: A higher purchase price than for a comparable conventional vehicle occur with equal probabilities.	One-Sample Chi-Square Test	.000	Reject the null hypothesis.
38	The categories of Q23b: The need to plug the vehicle in to recharge the battery occur with equal probabilities.	One-Sample Chi-Square Test	.000	Reject the null hypothesis.
39	The categories of Q23c:Concerns about limited access to plug-in locations occur with equal probabilities.	One-Sample Chi-Square Test	.000	Reject the null hypothesis.
40	The categories of Q23d:Availability of desirable vehicle size or style occur with equal probabilities.	One-Sample Chi-Square Test	.000	Reject the null hypothesis.
41	The categories of Q23e:Reliability of the vehicle occur with equal probabilities.	One-Sample Chi-Square Test	.000	Reject the null hypothesis.
42	The categories of Q23f:On-going maintenance and operative costs (including battery replacement) occur with equal probabilities.	One-Sample Chi-Square Test	.000	Reject the null hypothesis.
43	The categories of Q23g: The ability to carry heavy loads occur with equal probabilities.	One-Sample Chi-Square Test	.000	Reject the null hypothesis.
44	The categories of Q24a: Cars, minivans, vans, pickups, and SUVs are not an important source of air pollution anymore. occur with equal probabilities.	One-Sample Chi-Square Test	.000	Reject the null hypothesis.

Asymptotic significances are displayed. The significance level is .05.

Hypothesis Test Summary

	Null Hypothesis	Test	Sig.	Decision
45	The categories of Q24b: Government rules allow minivans, vans, pickups, and SUVs to pollute more than passenger cars, for every gallon of gas used. occur with equal probabilities.	One-Sample Chi-Square Test	.000	Reject the null hypothesis.
46	The categories of Q24c:Cars, minivans, vans, pickups, and SUVs are an important source of the greenhouse gases that many scientists believe are warming the earth's climate occur with equal probabilities.	One-Sample Chi-Square Test	.000	Reject the null hypothesis.
47	The categories of Q24d: Government rules require minivans, vans, pickups, and SUVs to meet the same miles-per-gallon standards as passenger cars. occur with equal probabilities.	One-Sample Chi-Square Test	.000	Reject the null hypothesis.
48	The categories of Q24e: Exhaust from cars, minivans, vans, pickups, and SUVs is an important source of the pollution that causes asthma and makes asthma attacks worse occur with equal probabilities.	One-Sample Chi-Square Test	.000	Reject the null hypothesis.
49	The categories of Q25: If very/somewhat interested in electric vehicles would you be willing to pay premium to purchase a plug-in hybrid vehicle? occur with equal probabilities.	One-Sample Chi-Square Test	.000	Reject the null hypothesis.
50	The categories of Age Group of the respodents occur with equal probabilities.	One-Sample Chi-Square Test	.000	Reject the null hypothesis.

Asymptotic significances are displayed. The significance level is .05.

There are differences between the observed frequencies 'Observed N' and the (theoretical) expected frequencies 'Expected N'. To perform the test we used the value of the A symp. Sig. which means asymptotic significance level for each of our research questions. It was found that it was statistically significant for all the variables as it was statistically significant at $p < 0.05$. For all other research questions, we will reject the hypothesis, which means that the sample distribution (observed frequencies) corresponds to the theoretical distribution (population).

Annexure 5 – Electric vehicle questionnaire

<table>
<tr><td>1</td><td colspan="4">Name (optional) ……………………………………… 1.a. Gender :</td></tr>
<tr><td>2</td><td>Age</td><td>a) 18- 30</td><td>b) 31 - 60</td><td>c) 61 or above</td></tr>
<tr><td>3</td><td>Education</td><td>a) Bachelor's</td><td>b) Master's</td><td>c) Student</td></tr>
<tr><td>4</td><td>Occupation</td><td>a) Service</td><td>b) Business</td><td>c) Student</td></tr>
<tr><td>5</td><td>Monthly income (INR)</td><td>a) 25,000 – 50,000</td><td>b) 50,000 – 100,000</td><td>c) Above 1L</td></tr>
<tr><td>6</td><td>Number of family members (tick all that apply)</td><td>a) 0 – 18 yrs.</td><td>b) 19 – 40 yrs.</td><td>c) 41 and above</td></tr>
<tr><td>7</td><td>Do you own a vehicle</td><td>a) Yes</td><td>b) No</td><td>c) Use someone's</td></tr>
<tr><td>8</td><td>Vehicle type</td><td>a) Car</td><td>b) Two wheeler</td><td>c) Other</td></tr>
<tr><td>9</td><td>Vehicle use</td><td>a) Personal</td><td>b) Business</td><td>c) Both</td></tr>
<tr><td>10</td><td>Vehicle name</td><td colspan="3"></td></tr>
<tr><td>11</td><td>No. of passengers</td><td colspan="3"></td></tr>
<tr><td>12</td><td>Drive per day</td><td colspan="3">a) Up to10 km b) 11 to 20 km c) 21 to 30 km
d) 31-60 km e) 60 km or more</td></tr>
<tr><td>13</td><td>How do you become aware of new technology?</td><td colspan="3">a) Television b) Newspaper c) Radio
d) Magazines e) Internet f) Word of mouth
g) Other</td></tr>
<tr><td>14</td><td>Who influences your vehicle purchase decision?</td><td colspan="3">a) Elders b) Children c) Joint decision</td></tr>
<tr><td>15</td><td colspan="4">When a new vehicle becomes available for purchase, what do you do?
a) I am among the first to purchase it.
b) I wait to read a review of it and then buy it if the review is favorable.
c) I wait until this new technology has been widely accepted and proven before considering it.
d) Other.</td></tr>
</table>

16. Please indicate how important each of the following considerations is to you in choosing the kind of vehicle you drive:

	Very Important	Important	Not Important
a) Fuel efficiency			
b) Safety			
c) Vehicle power			
d) Purchase price			
e) Reliability			
f) Vehicle size (to accommodate passengers or cargo)			
g) Expected operating costs (for maintenance and repair)			
h) Reputation of particular vehicle make or model			
i) Fuel type (e.g. petrol, diesel, CNG)			
j) Vehicle emissions and pollution			

17. How serious do you consider the following problems to be?

Please rate the problems as given below based on your perceptions

	Not a Problem	Problem	Major Problem
a) Traffic congestion that you experience while driving.			
b) Traffic noise that you hear at home, work or school.			
c) Vehicle emissions that affect local air quality.			
d) Vehicle emissions that contribute to climate change.			
f) Unsafe communities because of speeding traffic.			
g) Importing much of our oil from foreign countries			

18. Is your vehicle an electric vehicle?

    a) Yes b) No

19. Can you name any specific vehicle makes or models on the road today that are powered fully or in part by electricity?

    ………………………………………………………………………………………

    ………………………………………………………………………………………

20. How interested would you be in purchasing an electric motor vehicle once they become easily available in the next couple of years? Please tick.

    a) Very interested b) Somewhat interested
    c) Not very interested d) Not planning to purchase future vehicles
    e) Cannot say

21. Do you think transport policy of the State and Central Government should encourage electric vehicles?

    a) Yes, but only for public transport

    b) Yes, both for public and private transport

    c) No, not worth it

22. Please indicate how important each of the following reasons would be for you to consider purchasing or leasing an electric vehicle in the future

	Not Important	Important	Very Important
a) Saving money on the cost of operation (using electricity rather than gasoline)			
b) Reduced impact on the environment			
c) Reduced dependence on gasoline			
d) Driving a vehicle with more advanced or innovative technology			
e) Making a personal statement			

23. Please indicate how important each of the following reasons would be for you to not consider purchasing or leasing an electric vehicle in the future

	Not Important	Important	Very Important
a) A higher purchase price than for a comparable conventional vehicle			
b) The need to plug the vehicle in to recharge the battery			
c) Concerns about limited access to plug-in locations			
d) Availability of desirable vehicle size or style			
e) Reliability of the vehicle			
f) On-going maintenance and operative costs (including battery replacement)			
g) The ability to carry heavy loads			

24. This question deals with your views about the environmental impact of motor vehicles. For each of the following statements, please indicate if you agree or disagree. There is no correct answer.

	Agree	Disagree	Unsure
a) Cars, minivans, vans, pickups and SUVs are not an important source of air pollution anymore.			
b) Government rules allow minivans, vans, pickups and SUVs to pollute more than passenger cars, for every gallon of gas used.			
c) Cars, minivans, vans, pickups and SUVs are an important source of the greenhouse gases that many scientists believe are warming the earth's climate.			
d) Government rules require minivans, vans, pickups and SUVs to meet the same miles-per-gallon standards as passenger cars.			
e) Exhaust from cars, minivans, vans, pickups and SUVs is an important source of the pollution that causes asthma and makes asthma attacks worse			

25. If very/somewhat interested in electric vehicles would you be willing to pay a premium to purchase an electric vehicle?

a) 10% b) 15% c) 20%

References

1. "In a Nutshell – India's National Electric Mobility Mission Plan 2020." Http://indianautosblog.com/. 21 Jan. 2013. Web. 15 Sept. 2015. http://indianautosblog.com/2013/01/national-mission-for-electric-mobility-2020-61047

2. "National Electric Mobility Mission Plan." Http://pib.nic.in/. March 10, 2015. Accessed September 13, 2015. http://pib.nic.in/newsite/PrintRelease.aspx?relid=116719

3. Adler, Thomas, Laurie Wargelin, Lidia P. Kostyniuk, Chris Kavalec, and Gary Occhuizzo (2003). "Incentives for Alternate Fuel Vehicles: A Large-Scale Stated Preference Experiment," Paper presented at the 10th International Conference on Travel Behaviour Research, Lucerne, Switzerland, August 10-15.

4. Akerlof, G., R. Kranton (2000). Economics and Identity (Quarterly. Journal of Economics. 115 715–753.

5. Andan, D. and Faivre D'Arcier, B. (1997). Will the introduction of battery electric vehicles deeply change individuals' attitude towards car use? Conference Pre-print, Workshop on Response to New Transport Alternatives and Policies, International Association for Travel Behaviour Research, Austin, Texas, September.

6. Augustine, L.K., Sandidge S. A & Tim N. Williams (2011). Alternative Fuel Vehicles: Which Shall Win the Race to Commercialization? ISTM 6233 Emerging Technologies, George Washington University.

7. Ayre, J. (2015, May 6). Europe Electric Car Sales Booming. Retrieved September 15, 2015, from http://cleantechnica.com/2015/05/06/europe-electric-car-sales-booming/

8. Beggs, S., Cardell, S., & Hausman, J. (1981). Assessing the potential demand for electric cars. *Journal Of Econometrics*, *17*(1), 1-19. http://dx.doi.org/10.1016/0304-4076(81)90056-7

9. Bhakta, M. (2015). *Global Opportunities for SMEs in Electro-Mobility.* www.go4sem.eu. Retrieved from https://www.vdivde-it.de/eutool-go4sem/public/global-opportunities/india-2/at_download/file

10. Brownstone, D., Bunch, D. S., and Train, K. (2000). Joint mixed logit models of stated and revealed preferences for alternative-fuel vehicles. *Transportation Research* B, 34 (5), 315-338.

11. Bunch, David S., Bradley, M., Golob, T. F., Kitamura, R., and G.P. Occhiuzzo (1993). Demand for clean-fuel vehicles in California: A discrete-choice stated preference pilot project. *Transportation Research* A, 27 (3), 237-253.

12. California Energy Commission (CEC) (1999) *ABCs of AFVs: A Guide to Alternative Fuel Vehicles*, 5th Edition. http://www.energy.ca.gov/afvs/reports/1999-11_500-99-013.PDF.

13. Cao, X. (2004). THE FUTURE DEMAND FOR ALTERNATIVE FUEL PASSENGER VEHICLES: A DIFFUSION OF INNOVATION APPROACH. Sacramento, CA: The California Department of Transportation. Retrieved from http://www.tc.umn.edu/~cao/AQP_Cao.pdf

14. Caulfield, B., Farrell, S., & McMahon, B. (2010). Examining individuals preferences for hybrid electric and alternatively fuelled vehicles. *Transport Policy*, *17*(6), 381-387. http://dx.doi.org/10.1016/j.tranpol.2010.04.005

15. China Filling Station and Gas Station Industry Report,2015-2018. (2015, April 22). PRNewswire.

16. Dargay, Joyce and Gately, Dermot (1999). Income's effect on car and vehicle ownership, worldwide 1960-2015. *Transportation Research* A, 33 (2), 101-138.

17. David Brownstone, David S. Bunch, and Kenneth Train, "Joint Mixed Logit Models of Stated and Revealed Preferences for Alternative-Fuel Vehicles," *Transportation Research Part B* 34, no. 5 (2000): 315–338;

18. David S. Bunch, Mark Bradley, Thomas F. Golob, Ryuichi Kitamura, and Gareth P. Occhiuzzo, "Demand for Clean-Fuel Vehicles in California: A

Discrete-Choice Stated Preference Pilot Project," *Transportation Research Part A* 27, no. 3 (1993): 237–253;

19. Dimitris Potoglou and Pavlos S. Kanaroglou, "Household Demand and Willingness to Pay for Clean Vehicles," *Transportation Research Part D* 12, no. 4 (2007): 264–274.

20. *Driving Range for the Model S Family*. (2014). *Teslamotors.com*. Retrieved 9 March 2016, from https://www.teslamotors.com/blog/driving-range-model-s-family

21. Earthineer, (2016). Earthineer | Where homesteaders connect, learn, and trade. [online] Available at: http://www.earthineer.com/topic/48 [Accessed 10 Feb. 2016].

22. *Electric Auto Association*. (2016). *Electricauto.org*. Retrieved 11 March 2016, from http://www.electricauto.org/

23. Ewing, G., & Sarigöllü, E. (1998). Car fuel-type choice under travel demand management and economic incentives. *Transportation Research Part D: Transport And Environment*, *3*(6), 429-444. http://dx.doi.org/10.1016/s1361-9209(98)00019-4

24. Firstpost, (2015). FAME India: Govt scheme offers up to Rs 1.38 lakh incentives for electric, hybrid vehicles - Firstpost. [online] Available at: http://www.firstpost.com/business/fame-india-govt-scheme-offers-up-to-rs-1-38-lakh-incentives-for-electric-hybrid-vehicles-2189845.html [Accessed 18 Feb. 2016].

25. *Forbes Welcome*. (2016). *Forbes.com*. Retrieved 11 March 2016, from http://www.forbes.com/forbes/welcome/

26. Gallagher, K., & Muehlegger, 2011. E. Giving Green to Get Green: Incentives and Consumer Adoption of Hybrid Vehicle Technology. SSRN Electronic Journal. http://dx.doi.org/10.2139/ssrn.1083716.

27. Global Opportunities for Electric Mobility: India. (2015). [online] Available at: http://www.go4sem.eu/public/global-opportunities/china-1/electric-vehicle-supply-chain. [Accessed 18 Feb. 2016].

28. Golob, T. F., Torous, J., Bradley, M., Brownstone, D., Crane, S. S., and D.S. Bunch (1997). "Commercial fleet demand for alternative-fuel vehicles in California". Transportation Research A, 31 (3), 219-233.

29. Golob, Thomas F., Kitamura, Ryuichi Mark Bradley, and D.S.Bunch (1993). "Predicting the market penetration of electric and clean-fuel vehicles". *Science of the Total Environment* 134, no. 1-3: 371–381.

30. Gordon Ewing and Emine Sarigöllü, "Assessing Consumer Preferences for Clean-Fuel Vehicles: A Discrete Choice Experiment," *Journal of Public Policy & Marketing* 19, no. 1 (2000): 106–118;

31. Gordon O. Ewing and Emine Sarigöllü (1998). "Car fuel-type choice under travel demand management and economic Incentives". *Transportation Research Part D* 3, no. 6: 429–444.

32. Greene, D.L. (1996). "Survey evidence on the importance of fuel availability to the choice of alternative fuels and vehicles". *Energy Studies Review* 8, no. 3: 215–231.

33. Gulati, C. & Kandlikar, M. (2010). "Green drivers or free riders: an analysis of tax rebates for hybrid vehicles". *Journal of Environmental Economics and Management* 60, 78-93.

34. Hackbarth, A., & Madlener, R. Willingness-to-Pay for Alternative Fuel Vehicle Characteristics: A Stated Choice Study for Germany. SSRN Electronic Journal. http://dx.doi.org/10.2139/ssrn.2434523.

35. *High VAT in Certain States Killing Electric Vehicles Industry - The Automotive India*. (2014). *The Automotive India*. Retrieved 10 March 2016, from http://www.theautomotiveindia.com/high-vat-certain-states-killing-electric-vehicles-industry/

36. *History of the electric car | LosApos.com*. (2016). *Losapos.com*. Retrieved 11 March 2016, from http://www.losapos.com/electric_car_history

37. Kahn, Matthew E. (2006). "Do greens drive hummers or hybrids? Environmental ideology as a determinant of consumer choice, Institute of the Environment, UCLA, La Kretz Hall, Suite 300, Box 951496, Los Angeles, CA 90095, USA.

38. Kenneth Train, "The Potential Market for Non-Gasoline-Powered Vehicles," *Transportation Research Part A* 14, no. 5-6 (1980): 405-414;

39. Kenneth Train, Qualitative Choice Analysis: Theory, Econometrics, and an Application to Automobile Demand, Cambridge, MA: MIT Press, 1986.

40. Khan, A. M. and Willumsen, L. G. (1986). "Modeling car ownership and use in developing Countries". *Traffic Engineering and Control*, 27 (11), 554-560.

41. Madre, J. L. (1990). "Long term forecasting of car ownership and car use". In *Developments in Dynamic and Activity-based Approaches to Travel Analysis* (Jones, P., ed.), Avebury, UK, 406-416.

42. Mahindra Reva Electric Vehicles Pvt. Ltd,. (2013). *Mahindra Reva inaugurates electric vehicles charging station at Bengaluru International Airport*. Retrieved from http://www.mahindra.com/news-room/press-release/1385381759

43. Mokhtarian and Cao (2003). The Future Demand for Alternative Fuel Passenger Vehicles: A Preliminary Literature Review, Uc-Davis-Caltrans Air Quality Project- http://aqp.engr.ucdavis.edu, Task order No.31, Report July 13.

44. Naik, A. (2015, September 12). Top 5 Hybrid/Electric Cars in India. Retrieved September 16, 2015, from http://auto.ndtv.com/news/top-5-hybrid-electric-cars-in-india-757164.

45. Nixon, H., & Saphores, J. (2011). *Understanding household preferences for alternative-fuel vehicle technologies*. San Jose, CA: Mineta Transportation Institute, College of Business, San José State University.

46. Perkowski, J. (2014, December 12). Electric Cars: A Review Of 2014. Retrieved September 16, 2015, from http://www.forbes.com/sites/jackperkowski/2014/12/12/electric-cars-a-review-of-2014/

47. Pfaff, A., Chaudhuri, S., & Nye, H. (2004). Household Production and Environmental Kuznets Curves – Examining the Desirability and

Feasibility of Substitution. *Environmental And Resource Economics*, *27*(2), 187-200. http://dx.doi.org/10.1023/b:eare.0000017279.79445.72

48. Potoglou, Dimitris and Pavlos S. Kanaroglou (2008). "Disaggregate demand analyses for conventional and alternative fueled automobiles: A review". *International Journal of Sustainable Transportation* 2, no. 4: 234-259.

49. Prakash, N., Kapoor, R., Kapoor, A., & Malik, Y. (2014). GENDER PREFERENCES FOR ALTERNATIVE ENERGY TRANSPORT WITH FOCUS ON ELECTRIC VEHICLE. *Journal Of Social Sciences*, *10*(3), 114-122. http://dx.doi.org/10.3844/jssp.2014.114.122

50. Ramadhas, S. Arumugam (2011). Alternative Fuels for Transportation, Taylor & Francis, London.

51. Rodrigue, J. & Notteboom, T. (2016). Transportation and Economic Development. People.hofstra.edu. Retrieved 8 Feb 2016, from http://people.hofstra.edu/geotrans/eng/ch7en/conc7en/ch7c1en.html

52. Schulze, Tim, BeateMüller, and Gereon Meyer. Advanced Microsystems For Automotive Applications 2015. Print.

53. Shahan, Zachary. "New Electric Cars In 2015 — I'M Counting 15". CleanTechnica. N.p., 2014. Web. 1 March 2016.

54. *Short History of Electric Vehicles - Anandalal Electric*. (2016). *Anandalalelectric.com*. Retrieved 11 March 2016, from http://www.anandalalelectric.com/short-history-of-electric-vehicles

55. SUNDAS, S. (2015, June 9). On the road to zero emissions, Bhutan hits a few electric car bumps. Retrieved September 13, 2015, from http://www.reuters.com/article/2015/06/09/us-bhutan-climate-change-autosales-idUSKBN0OP0W720150609

56. T.S. Turrentine and Kurani, K.S. (2007). "Car buyers and fuel economy"; *Energy Policy* 35 1213–1223

57. Thomas Adler, Laurie Wargelin, Lidia P. Kostyniuk, Chris Kavalec, and Gary Occhuizzo, "Incentives for Alternate Fuel Vehicles: A Large-Scale

Stated Preference Experiment," Paper presented at the 10th International Conference on Travel Behaviour Research, Lucerne, Switzerland, August 10-15, 2003.

58. Train, K. (1986). *Qualitative Choice Analysis: Theory, Econometrics, and an Application to Automobile Demand (MIT Press series in transportation studies; 10)*. MIT Press.

59. TSHERING, L. (2014, November 1). BHUTAN ELECTRIC VEHICLE INITIATIVE: A STEP TOWARDS ZERO EMISSION. Retrieved September 15, 2015, from http://www.uncrd.or.jp/content/documents/22548EST-P2_Bhutan.pdf

60. Turrentine, T., & Kurani, K. (2007). Car buyers and fuel economy? *Energy Policy*, *35*(2), 1213-1223. http://dx.doi.org/10.1016/j.enpol.2006.03.005

61. Voelcker, J. (2014, July 29). 1.2 Billion Vehicles On World's Roads Now, 2 Billion By 2035: Report.

62. Vorrath, S. (2015, August 17). Five things you didn't know about the electric vehicle market. Retrieved September 12, 2015, from http://reneweconomy.com.au/2015/five-things-you-didnt-know-about-the-electric-vehicle-market-97757

63. Wessel, N. (2012). What is transportation? Why do we need it? *cincymap*. Retrieved from http://cincymap.org/blog/what-is-transportation/

64. Ziegler, A. (2012). Individual characteristics and stated preferences for alternative energy sources and propulsion technologies in vehicles: A discrete choice analysis for Germany. *Transportation Research Part A: Policy And Practice*, *46*(8), 1372-1385. http://dx.doi.org/10.1016/j.tra.2012.05.016

65. Sparke, L. Reducing Emissions Associated with Electric Vehicles (1st ed., p. 2). EDay Life Ltd. Retrieved from http://www.edaylife.com.au/library/EV%20Emissions.pdf

www.ingramcontent.com/pod-product-compliance
Ingram Content Group UK Ltd.
Pitfield, Milton Keynes, MK11 3LW, UK
UKHW041637190726
13854UKWH00006B/2541

9 789385 020711